Studies in Big Data

Volume 172

Series Editor

Janusz Kacprzyk, Polish Academy of Sciences, Warsaw, Poland

The series "Studies in Big Data" (SBD) publishes new developments and advances in the various areas of Big Data- quickly and with a high quality. The intent is to cover the theory, research, development, and applications of Big Data, as embedded in the fields of engineering, computer science, physics, economics and life sciences. The books of the series refer to the analysis and understanding of large, complex, and/or distributed data sets generated from recent digital sources coming from sensors or other physical instruments as well as simulations, crowd sourcing, social networks or other internet transactions, such as emails or video click streams and other. The series contains monographs, lecture notes and edited volumes in Big Data spanning the areas of computational intelligence including neural networks, evolutionary computation, soft computing, fuzzy systems, as well as artificial intelligence, data mining, modern statistics and Operations research, as well as self-organizing systems. Of particular value to both the contributors and the readership are the short publication timeframe and the world-wide distribution, which enable both wide and rapid dissemination of research output.

The books of this series are reviewed in a single blind peer review process.

Indexed by SCOPUS, EI Compendex, SCIMAGO and zbMATH.

All books published in the series are submitted for consideration in Web of Science.

Gokhan ALTAN · Ipek ABASIKELEŞ TURGUT
Editors

Deep Learning in Ad-Hoc Wireless Networks

Springer

Editors
Gokhan ALTAN
Department of Computer Engineering
Iskenderun Technical University
Hatay, Türkiye

Ipek ABASIKELEŞ TURGUT
Department of Computer Engineering
Iskenderun Technical University
Hatay, Türkiye

ISSN 2197-6503 ISSN 2197-6511 (electronic)
Studies in Big Data
ISBN 978-3-031-86074-4 ISBN 978-3-031-86075-1 (eBook)
https://doi.org/10.1007/978-3-031-86075-1

This Springer imprint is published by the registered company Springer Nature Switzerland AG
The registered company address is: Gewerbestrasse 11, 6330 Cham, Switzerland

Contents

Recent Deep Learning Based Trust Solutions

Ipek ABASIKELEŞ TURGUT and Gokhan ALTAN

Abstract The vulnerability of wireless communication to attacks, combined with the highly dynamic nature of VANETs and the stringent requirements for accuracy and speed, necessitates the development of specialized solutions for these networks. Trust systems have emerged as an important area of interest in both academic and industrial contexts, especially considering the limitations of cryptographic methods that primarily address external threats. While trust systems utilizing machine learning have been used for some time, the application of deep learning in this field is relatively new. This study presents an analysis of the algorithms used and their application areas, focusing on recent developments in the integration of deep learning approaches within trust systems. It is anticipated that the findings of this study will contribute to the future development of deep learning-based trust systems.

Keywords Deep learning · VANET · CNN · Review · DNN · ML · DBN

1 Introduction

In recent years, increasing globalization and population have caused a significant increase in the number of travelers. In addition, rapid developments in the automotive sector have enabled the production of cars suitable for every budget, which has supported the increase in the number of personal vehicles. The increase in the number of individual vehicles in traffic has negative effects. The most obvious of these is traffic congestion. Especially in metropolitan areas, during work and exit hours, or in adverse weather conditions, both congestion and the number of accidents increase.

These authors contributed equally to this work.

I. ABASIKELEŞ TURGUT · G. ALTAN
Computer Engineering Department, Iskenderun Technical University, Hatay, Türkiye
e-mail: ipek.abasikeles@iste.edu.tr

G. ALTAN (✉)
R&D Department, Deeptech Engineering Trade Co. Ltd., İskenderun, Hatay, Türkiye
e-mail: gokhan.altan@iste.edu.tr

G. ALTAN and I. ABASIKELEŞ TURGUT (eds.), *Deep Learning in Ad-Hoc Wireless Networks*, Studies in Big Data 172,
https://doi.org/10.1007/978-3-031-86075-1_1

This situation not only reduces the efficiency of traffic but also endangers the safety of drivers and passengers [1]. According to TUİK data [2], a total of 1 million 314 thousand 136 traffic accidents occurred on the Turkish highway network in 2023. 1 million 79 thousand 65 of these accidents involved material damage and 235 thousand 71 involved fatal injuries. Considering this report, while ensuring the transportation safety of vehicles with smart traffic management systems, increasing traffic efficiency has become a focus of attention in both academia and industry in recent years.

Intelligent transportation systems utilize sensors embedded in modern vehicles to effectively manage traffic. These sensors, which perceive various environmental and traffic conditions, exchange information with their neighbors by forming wireless networks. These networks are called VANETs. VANETs are a specialized type of MANETs where the nodes are vehicles. The basic elements of VANET networks are, as seen in Fig. 1, the vehicles equipped with OBUs that perform information exchange and AUs that run applications, and the RSUs that provide interaction between the vehicles and the global Internet, located at the roadside [3].

In VANETs, vehicles can communicate with each other, with sensors, with passengers, and with infrastructures. To describe all of these communications, V2X vehicletoeverything (Fig. 2) is used, while V2V is used for the communication between vehicles [4]. Thanks to these communications, different types of services are provided to drivers, passengers, and pedestrians. These services are divided into two main categories: safety and non-safety applications [5]. Safety applications include safety-related messages such as accidents and/or collisions, dangerous road conditions, traffic congestion, roadworks, emergency vehicle approaches, emergency

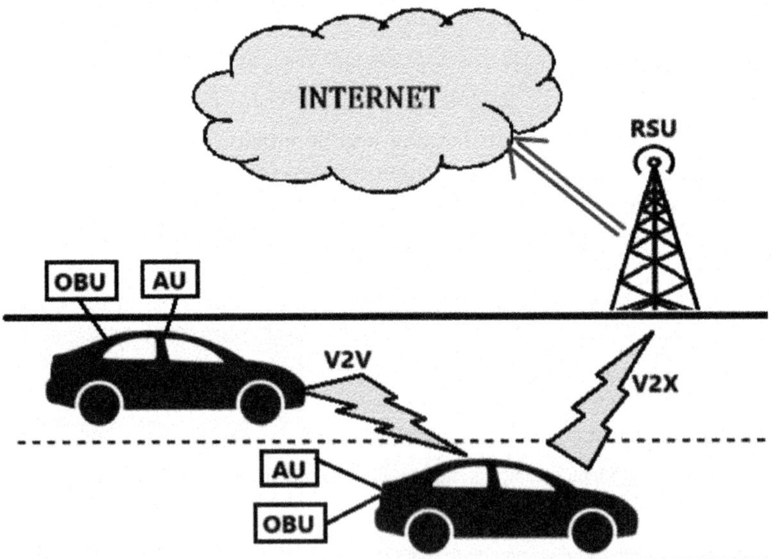

Fig. 1 General structure of VANET

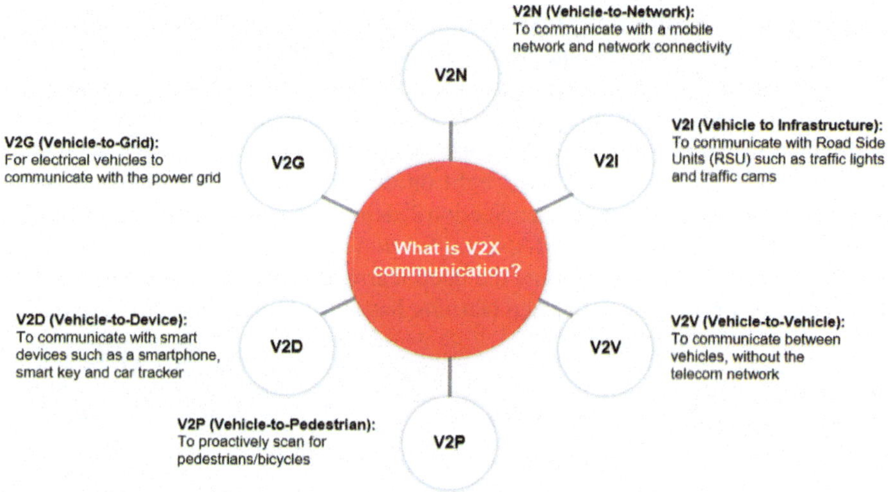

Fig. 2 The communication environment of V2X [7]

braking and sudden deceleration, etc., as well as notifications and warnings. The large number of messages generated by these applications must be not only transmitted quickly in real time but also accurately. The open nature of the wireless environment can endanger these networks. Therefore, security and privacy issues must be completely resolved before VANET applications can be implemented in practice [6].

Traditional security solutions cannot be directly applied to these networks due to the high mobility, low delay constraints, and dynamic environment characteristics specific to VANETs [8]. This situation has led to the development of solutions specific to these networks. These solutions are divided into two main categories cryptological approaches and trust-based mechanisms [9]. Cryptological solutions provide authentication of the vehicles using digital signatures, and public and private key pairs and thus protect the network from unauthorized access. However, it cannot take any precautions against the injecting of false messages by authorized vehicles with keys, whether intentionally or unintentionally, and the problems this situation will create. In recent years, trust mechanisms have attracted attention in both academia and industry as an attractive solution for this group of attacks called internal attacks.

VANETs strive to improve the safety and functionality of transportation systems by enabling communication among vehicles, pedestrians, and roadside units. Artificial intelligence (AI) methods, recognized for their superior ability to address complex problems, are now being utilized across a variety of real-world scenarios. The advancement of computationally efficient techniques and the abundance of large-scale data have been key drivers of AI's growing success. Both machine learning (ML) and deep learning (DL) have experienced notable progress in recent years, shifting from theoretical exploration in controlled environments to real-world automation with significant and diverse applications [10].

In this study, the use of deep learning approaches on VANET trust systems has been investigated. For this purpose, 10 recent approaches in literature are examined and which techniques are used for what purpose in each of them are detailed. It is thought that this study will shed light on deep learning-based trust approaches to be implemented in the future. Section 2 demonstrates the unique characteristics of VANET different from MANETs. Section 3 examines the recent trust-based approaches, while Sect. 4 explains the deep learning approaches used in VANETs. Finally, Sect. 4.6 evaluates the deep learning based trust mechanism in VANETs.

2 Characteristics of VANET

Trust management schemes have long been a preferred method in non-VANET networks (especially in MANETs). The reason why these approaches cannot be directly applied to VANETs is that they do not provide an optimum solution due to the unique characteristics of VANETs (Fig. 3). The main features that distinguish VANETs from MANETs are listed below.

1 **High speed**: One of the most important characteristics of VANET networks is that network elements (vehicles) have high speeds. Network elements with speeds of up to 200 km/h for the VANET highway scenario are not usually seen in MANET

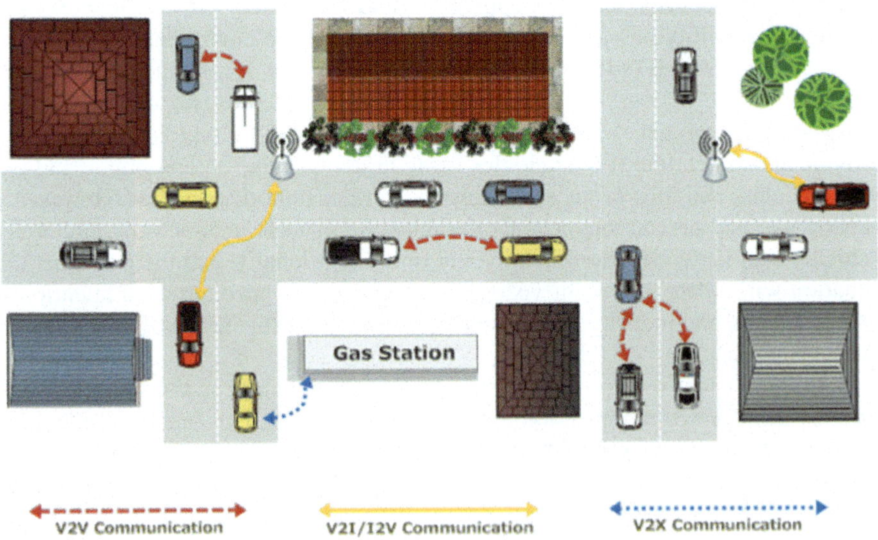

Fig. 3 Characteristics of VANET [11]

scenarios. This makes both the creation of the network topology and the maintenance of the created topology quite difficult.

2 **Delay constraint**: With high speeds, there is a very strict constraint on the delay of messages, especially for safety applications. Although there is no strict delay constraint in infotainment applications, there is a high data throughput demand. Therefore, the solutions proposed for VANET must meet the delay constraint of safety applications and the high throughput demand of entertainment applications together.
3 **Resource availability**: Generally, the elements that makeup MANET networks are elements with limited resources such as processors, batteries, and communication distance. Therefore, one of the main goals of MANET solutions is to use energy and storage effectively and to minimize the need for processing with lightweight algorithms. The absence of these constraints in the vehicles that make up VANETs allows the use of more powerful algorithms.
4 **Limited mobility**: While MANET elements can mostly move unlimitedly along a 360-degree angle, in VANET networks the direction of movement is limited by the road topology. Especially in highway scenarios, vehicles mostly move in one direction. Therefore, the solutions required for these networks do not need to make additional efforts to provide mobility.
5 **Rapidly changing topology**: Due to high speeds, there is a very frequently changing topology in the network. For example, cluster approaches can be managed more easily in MANETs because they change more slowly, but in VANETs, rapid topological changes require constant updating. Therefore, the solutions to be proposed in these networks must keep up with these topological changes and be constantly updated.

3 Trust Mechanisms in VANET

Trust management methods are a preferred method in the literature to protect various types of networks [4]. Trust, in its most basic form, expresses the probability that a node in the network is honest and takes a value between 0 and 1. The value 0 is used to indicate that it is completely untrustworthy, while the value 1 is used to indicate that it is completely trustworthy. If this value is lower than a certain threshold, the evaluated node is labeled as an attacker. There are two types of elements in the system: the evaluator and the evaluated. Evaluator nodes are the nodes that determine the trust value for the evaluation. There are many factors to be preferred in the design of the trust management system developed for this evaluation. Design metrics are shown in Fig. 4.

There are two approaches to the trust management scheme architecture: distributed and centralized. In the centralized approach, a device is generally used in the role of a trusted authority to calculate the trust of the vehicles globally and distribute these values to the vehicles in the network. In this approach, more accurate evalu-

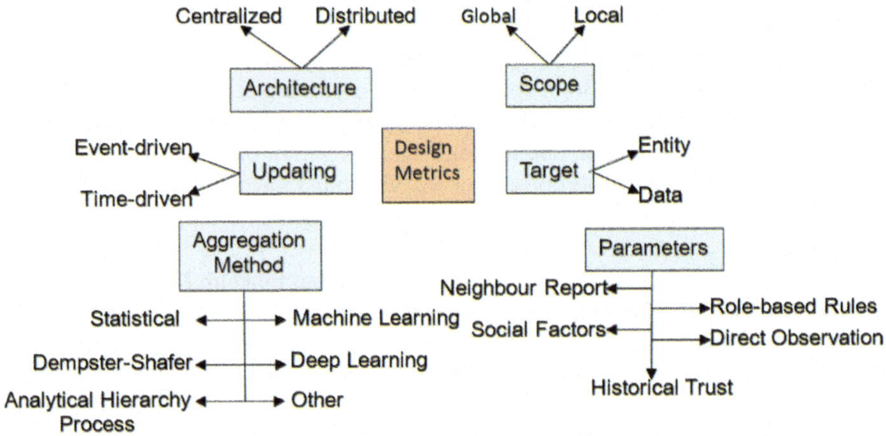

Fig. 4 Trust management system design metrics

ations (i.e. lower false positive rates) can be made thanks to the global perspective. However, incomplete evaluation against blind spots and the lack of central authority devices, especially in rural areas, limit the scalability of this approach. In distributed approaches, trust values are calculated locally by the devices completely independently. Vehicles communicate with their environment to make more accurate trust evaluations. In this approach, the blind spot problem caused by the lack of central architecture is not experienced. However, due to the lack of global information, the probability of vehicles making false positive evaluations increases. In recent years, architectures combining both approaches have been proposed in the literature [12]. In points where the lack of a central structure is felt, the distributed approach, and the centralized approach are used to solve the scaling problem and reduce the false positive rates.

Trust calculation location is divided into two: global and local. Local trust value is an evaluation made independently from a central structure within the vehicles. Since the environment, location, and status of the vehicle are important in this evaluation, it does not reflect the general status of the system. The local trust value calculated for the same evaluated vehicle may be different from each other in each evaluator vehicle. Global trust is 1 for each vehicle in the network and reflects the general status of the network. A central device/element is used for global trust calculation. Trust is divided into two groups according to its target: entity and data. The trust in the vehicle is used in calculating entity trust. If a vehicle is labeled as unreliable, all messages coming from that vehicle are evaluated as unreliable. Similarly, all messages coming from a reliable vehicle are also labeled as reliable. Due to environmental conditions, messages from reliable vehicles may also contain errors/mistakes in some cases. In this case, vehicle reliability alone is insufficient in trust assessment. For this reason, in data trust, the reliability of the message, not the vehicle, is evaluated. Today, many approaches use both entity and data trust together [13–16].

After calculating trust values in the network, whether local or global, an update is needed. The update approach can be triggered depending on the occurrence of a specific event in the network, or it can be done periodically. In global trust calculations, time-based triggering is usually used. Periodic updates can be fixed-frequency [15], or variable-frequency [12] depending on the trust status of the network. In local trust calculations, an event-dependent triggering approach is usually used [12, 15].

The parameters used in trust calculation vary in many studies. In general, trust assessment metrics are divided into two: direct and indirect observation. Direct observation includes the experiences of the assessor vehicle itself for the target range. Indirect observation reflects the information obtained from the assessor vehicle's neighbors. If the assessor and the assessed vehicle are not in each other's coverage area, indirect observation values are used. Apart from these, fixed trust values assigned depending on the roles of the vehicles are also used. For example, assigning high trust values to official vehicles, state-affiliated and special-duty vehicles (e.g. ambulances, police cars, fire trucks, etc.) is a frequently used approach [12, 13, 17]. In the periodic calculation of the global trust value, the previous trust value is usually used. Thus, the vehicle's tendency (for better or worse) can be followed more closely. Finally, some studies [18] have included social factors such as the vehicle's age, the driver's accident history, etc. in the trust calculation. The parameters to be selected in the trust calculation, the weights they have, or how they will affect them vary across studies. The general approach is to combine the parameters under a function using statistical methods under fixed weights. Apart from this, there are also approaches such as machine learning, deep learning, Dempster-Shafer Theory, Analytical Hierarchy Process. This article focuses on how deep learning methods are used in trust calculation.

4 Deep Learning in VANETs

Although machine learning algorithms have long been used in VANET, in recent years, deep learning techniques have also gained attention in academic and industrial societies. Machine learning has been used in safety, communication, and traffic-related issues, including traffic flow, congestion, scheduling, and monitoring, in VANETS [19]. For instance, driver assistance, collision alert systems, and vehicle safety are some of the usage areas of safety applications in VANETs, in which machine learning methods have been conducted. Similarly, these techniques have also been utilized in data congestion control, misbehavior detection, link management, and routing issues.

Recent deep learning approaches used in trust mechanisms in VANETs include Deep Neural Networks (DNN) [1, 20], Deep Belief Networks (DBN) [21], Recurrent Neural Networks (RNN) [22], Convolutional Neural Networks (CNN) [23], and Deep Maxout Networks (DMN) [24].

4.1 Deep Neural Networks (DNN)

Multilayer Perceptron (MLP), one of the most popular machine learning algorithms, has maintained its popularity for years with nonlinear inference and bakpropagation procedure for the learning process on the input data. The increase in features based on different domains has led to the need to analyze the focused problem in more detail, especially for big data. Deep Neural Networks (DNNs) have gained great popularity in many recent researching areas. DNNs are types of artificial neural networks with many layers and big number of neurons at each layer (see Fig. 5). It means that the number of hidden layers between the input and output layers is more than one. Compared to shallow MLP architectures, modeling deep architectures allows learning more complex data and prediction processes with higher generalization capacity for even real and big data. Contrarily, it also increases the number of important training hyper parameters such as the number of layers, units per layer, weights, and more. It consists of many layers and these layers connect neurons to solve complex activation functions [25].

DNN is an enhanced type of artificial neural network that can be adapted to classification, regression, and generalization processes. It emerged with the idea of increasing the number of neurons in each layer, just as it increases the number of layers in order to model the data association among neurons. We can think of deep learning as a generalization of a human being in the sense that it has several layers of human neuron cells. Just as a human brain can recognize a certain image, time series, 3D prediction and distinguish it from another inputs, a DNN has ability to learn a model that can recognize and classify similar inputs. Like the connection and electrical activity in human brain, DNN utilizes a vector, which is an artificial

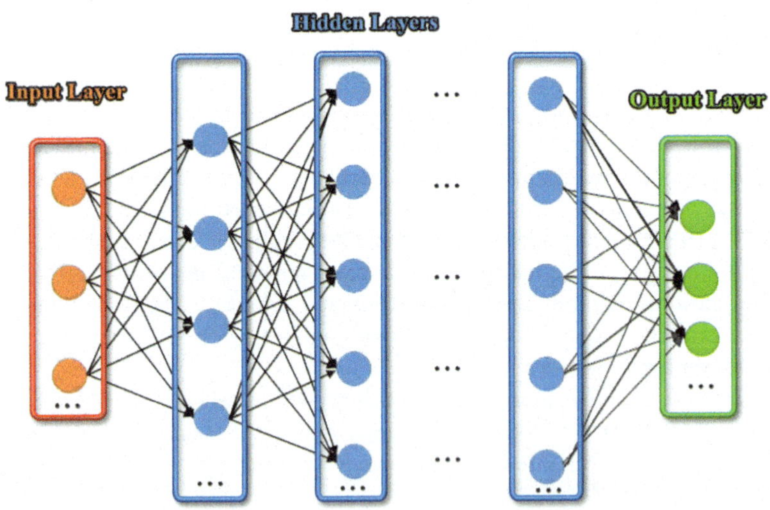

Fig. 5 DNN architecture

sensory of input [26]. It provides the input in the form of a matrix called the input layer by vectorization often with various feature extraction methods depending on the type of data or as the output of unsupervised models. DNN algorithm is performed a crucial role as the learning model of many deep learning approaches [25, 26]. With the integration of fully connected layer and different optimization, standardization, and regularization techniques, it positively improves the train and prediction performances of the system in parallel with the amount of data.

4.2 Deep Belief Networks (DBN)

Deep belief network (DBN) is a type of semi-supervised learning with more hidden layers and units per layer as a traditional and practical deep learning approach (see Fig. 6). It is a probabilistic graphical nonlinear method that utilizes multilayer nonlinear mappings to learn various distributions and representations for complex data using different models. Feature representation is performed by the network's features using unsupervised algorithms [27]. Selecting the connection weights and model bias term parameters uses the restrictive bias technique during training of DBN model. Gibbs sampling is applied on the training data to update the pre-trained network parameters at each iteration of restricted Boltzmann machines (RBM) [28]. In the last stage of the DBN network, a supervised feedforward backpropagation neural network is adapted to update the model parameters on the network settings until the system error is less than the preferred error. DBN consists of a two-stage training [29]. The first stage

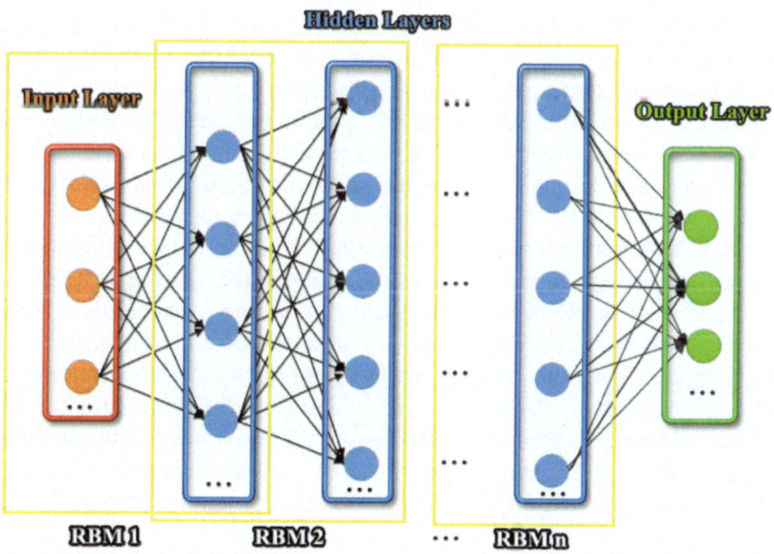

Fig. 6 DBN architecture

obtains the pre-trained weights with the probabilistic and energy-based functions. Thus, the pre-trained weights directly represent the input data using unsupervised. The pre-training stage with unsupervised learning is performed with RBM modules, which represents independently as a feature detector [27, 29].

DBN can effectively learn complex dependencies between variables. Since it contains the depth of architecture for many layers of hidden variables, it can effectively learn the intrinsic feature representations of the data. It can also be used as a nonlinear dimensionality reduction method with RBM training. DBN solves the optimization problem of DNN using layer-by-layer training [29]. The most important role in layer-wise training is undertaken by the RBM. DBN performs restrictive bias algorithm. DBN unfolds the pre-trained weights for a positive phase and a negative phase. The positive phase is utilized to calculate the activation probabilities of the hidden units by sampling the activation of the visible layer, and vice versa for the negative phase [28]. The adaptability of two-stage training on DBNs enables to effectively learn complex representations of the data, which makes them extremely useful, especially for tasks where the input data is high-dimensional and requires deep understanding.

4.3 Recurrent Neural Networks (RNN)

Recurrent neural network (RNN) is an advanced type of traditional feedforward neural network that can process a series of inputs of variable length. It has been included in the literature as a method designed specifically for modeling sequential data and time series. The reason why RNN is popular on time series is that it has a recurrent hidden state whose activation at each time depends on the previous outputs of the adjacent recurrent unit. Long short-term memory units (LSTM), which are gaining popularity, allow each recurrent unit (see Fig. 7) to adaptively capture the dependencies of different time scales using the main concept of RNN [30].

Time series contains intrinsic spatial information that cannot be transferred by an MLP. Whereas a simple neural network is processing instantaneous data, RNNs are reinforced with an additional time step unit with memory that brings a sense of time to the neural network [30]. The edges, called recursive units, allow a neuron to pass the output of its output unit to the next neuron in addition to its current input value. The association among adjacent units forms a loop in the training of RNN [31].

The capacity of an RNN is to manage and store the recurrent data points in a network cycle. In short, the RNN algorithm unfolds the logic from past interactions and the outputs of previous units to predict future events. RNNs can sort vectors and allow more complex processes to be undertaken. RNNs represent a series of interconnected networks while also transferring the relationship between these units with memory units [31]. Therefore, it aimed to complete this process with the use of a hidden layer with tanh and sigmoid activation functions. The main and most important concept of RNN is the hidden state that holds some information about a sequential information. Consequently, the output from the previous RNN unit is

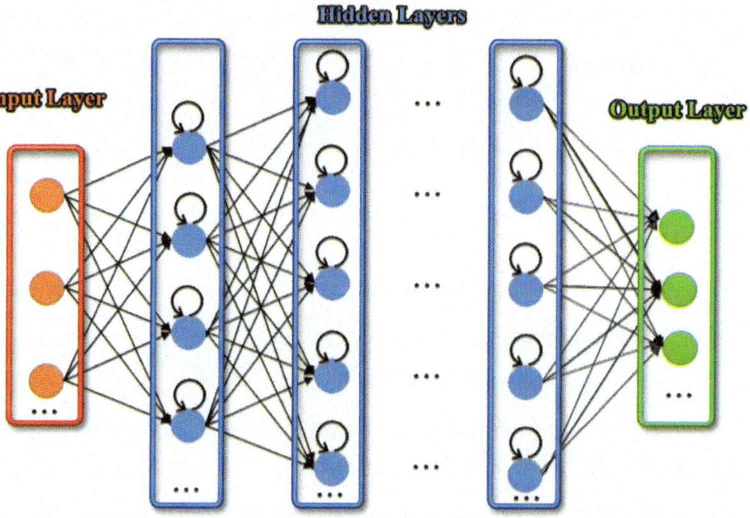

Fig. 7 Deep RNN architecture

compiled in relation to each other and transferred to the next units, allowing the model to learn dynamically from the past [30, 31]. Thanks to this specification, it is the main idea for the algorithms, such as LSTM, Bi-LSTM, GRU, and Bi-GRU.

4.4 Convolutional Neural Networks (CNN)

Convolutional neural network is the most widely used deep learning model for feature learning on large-scale image classification, detection and localization processes. It is maturely process a variety of data, including images, sounds, time-series, 3D images, and text to recognize specific features of datasets such as intensity, morphology, color-based features [32]. CNN discovers distributed feature representations of data by combining low-level features extracted in its early stages to obtain abstract high-level representations of mid- and high-level feature maps or features (see Fig. 8). Currently, it has been adopted into the most important stages of computer vision activities to solve problems in this area. In order to facilitate a full understanding of the CNN architecture, it is likely that the convolution process, which is an advanced image processing technique that gives its name to the CNN algorithm [33].

CNNs are also known as shift-invariant or space-invariant artificial neural networks based on a shared and pre-trained weight architecture of convolution kernels or filters. The filters slide along the input features and provide translation-covariant responses known as feature activation maps [34]. Therefore, CNNs has capabilities to represent spatial information of the input data. CNNs perform characterization of universal nonlinear function approximations. A CNN is comprised of three main

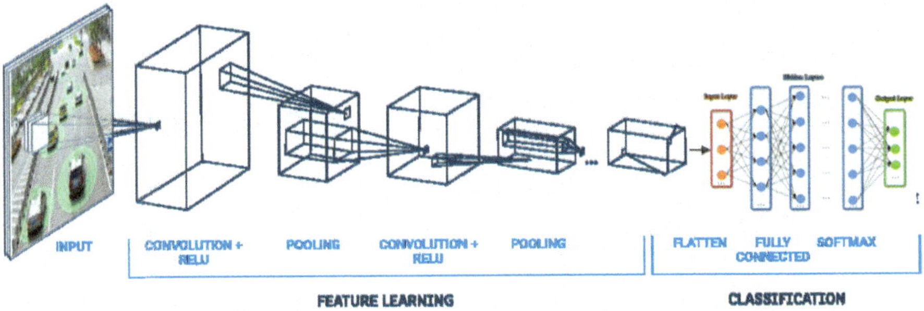

Fig. 8 Deep CNN architecture

layers, namely, convolutional layer, subsampling layer (pooling layer), and fully connected layer [26, 34, 35].

- Convolutional layer: It consists of several convolutional units that are given a certain size of square matrix called filter or kernel during the feature learning stage. The kernel is moved one step at a time to weigh the input data and extract different characteristics and representations. In short, this layer is the most basic building block of a feature learning stage. The size, number, frequency, and sequential of the filters allow different flatness increase the representational capability on the convolved features [33].
- Pooling layer: It is commonly placed in the middle of consecutive convolutional layers and is essentially a down-sampling layer used to compress the size of input data. It also reduces the number of the parameters and overfitting. The main role of the pooling layer is to compress the input by max and average down-sampling approaches. Pooling layer ensures circulating the most dominant region values among the subsequent layers [35].
- Batch normalization layer: It is proposed to meet the regularization difficulty of deep model training. Batch normalization calculates the mean and standard deviation in small groups to continuously adjust the intermediate output of the neural network. It makes the value of the intermediate output of the entire neural network in each layer constant and improves the convergence speed in the training [34].
- Excitation function layer: Its role in the training is to process the non-linearity of the data. It applies a corrected linear unit excitation function [26, 32, 33].
- Fully connected layer: The role of this layer is the stage used for the learning process of the dominant feature maps determined after the feature learning stages. It maps the learned distributed feature representations to the sample labeling space, which is simply represents the classification stage. The fully connected layer adapted used in many popular deep learning algorithms actually completes the process by taking different optimization methods and convergence approaches of the DNN structure as a basis [26, 33–35].

4.5 Deep Maxout Networks (DMN)

DMN is a type of DNN model with a sequentially combined multilayer architecture (see Fig. 9). DMN effectively inherits the advantages of a trainable and non-saturated activation function by prioritizing max concepts of the calculated unit. It is also effective in providing solutions in resource-constrained situations in terms of model and data. Each maxout layers in the DMN network contains hidden units with non-intersecting groups. It utilizes maxout functions to generate hidden activations of the units [36].

Maxout neuron is also a generalization type of ReLU. For Re-LU nonlinearity, max-pooling occurs between a single feature map and 0, which acts like a k-part maxout neuron. In contritely, maxout units perform feature selection intelligently, whereas ReLU only defines the value of the input unit. This may seem surprisingly simple compared to traditional neural network nonlinear functions such as logistic sigmoid or hyperbolic tangent. However, although its time-complexity is low against to these activation functions, it is capable of giving effective convergence capabilities on the training of the units. The main characteristics to define the competitiveness of the DMN are enabling the robust feature selection; proposing a generalization of ReLU; and defining the nonlinearity as a universal approximator [36].

The performance of a deep learning-based model is evaluated using the statistical test characteristics of the training and testing. Whereas the validation performance is important to lead the training accuracy for generalization, overfitting, and underfitting, the test metrics are utilized to evaluate the performance of the trained deep learning model for prediction over non-trained exclusive data [26, 32, 37]. The statistical metrics for training and testing are calculated using the contingency table,

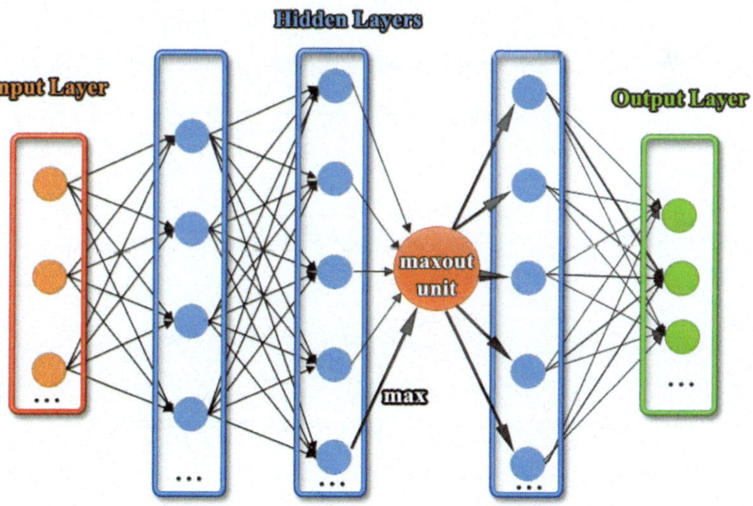

Fig. 9 Deep maxout network architecture

Table 1 Confusion matrix

		Predicted	
		Class 1	*Class2*
Actual	*Class 1*	TP	FN
	Class 2	FP	TN

which is a matrix representing the actual and predicted classes in machine learning. Therefore, overall accuracy (Eq. 1), sensitivity (recall) (Eq. 2), specificity (Eq. 3), precision (Eq. 4), f1-score (Eq. 5), and Matthews correlation coefficient (MCC) (Eq. 6 are calculated in the comparison of the deep learning architectures and algorithms.

Confusion Matrix (see Table 1) consists of four basic components:

- True Positive (TP) represents for the number of correctly predicted samples as "Class 1".
- True Negative (TN) is the number of correctly predicted samples as "Class 2".
- False Positive (FP) is the number of samples that are predicted as "Class 2" when it is actually "Class 1".
- False Negative (FN) is the number of incorrectly predicted samples as "Class 1" when it is actually "Class 2"

$$Overall\ Accuracy = \frac{TP + TN}{(TP + FN + FP + TP)} \tag{1}$$

$$Recall\ (Sensitivity) = \frac{TP}{(TP + FN)} \tag{2}$$

$$Specificity = \frac{TN}{(TN + FP)} \tag{3}$$

$$Precision = \frac{TP}{(TP + FP)} \tag{4}$$

$$F1 - score = \frac{2 \times TP}{(2 \times TP + FP + FN)} \tag{5}$$

$$MCC = \frac{TP \times TN - FP \times FN}{\sqrt{(TP + FP) \times (TP + FN) \times (TN + FP) \times (TN + FN)}} \tag{6}$$

4.6 Deep Learning Based Trust Approaches A Literature Review

Wang et al. [1] proposed a trust management scheme for the Internet of Vehicles using blockchain and deep learning. In the scheme they proposed, vehicles report the information they collect from their environment to the nearest RSU. They proposed a deep learning-based verification model for attacks that may occur in these reports (i.e. false message notification). This model is used both to calculate the reliability of messages and to obtain the credibility of vehicles. In the approach used, a FCN is created. In this network, 3 hidden layers are used in addition to the input and output layers. The first 2 layers contain 100 neurons, and the last one contains 50 neurons. FCN takes as input the time the vehicle uploads its report to neighboring RSU with the location, the role, familiarity, credibility, and speed of the vehicle, the location, and the type of event. The output of FCN is the trust value of the message. They evaluate their performance by simulation using a simulator of urban mobility (SUMO) for a real-time map of Manhattan from OpenStreetMap. The evaluation parameters include precision and recall. The simulation results show that the proposed approach can successfully evaluate the trust of nodes and detect the attacker vehicles.

Banguia et al. [38] tackle the issue of bias in managing indirect trust in digital settings. They introduce a reputation model that combines deep learning with synthetic data to improve decision-making and reduce biases. A deep generative model is used to generate synthetic data. For investigating the effect of generated synthetic data on trust classification, various classifiers including KNN, MLP, Decision Tree, Random Forest and SVM are utilized. Evaluation metrics are recall, F1-score, precision, accuracy, and Kappa. Findings reveal that using synthetic data markedly boosts classification accuracy, promoting fairer trust evaluations that consider both sentiment-based and neutral feedback.

Tangade et al. [20] introduces a deep learning-based scheme (DL-DCTC) for trust management in VANETs, addressing vulnerabilities to malicious nodes and internal attacks. Unlike cryptography-based methods, which can't prevent false information from legitimate senders, DL-DCTC uses neural networks to evaluate driver behavior and classify messages as trustworthy or not. In this paper, two sequential DNNs are proposed. The first model includes three hidden layers having 512, 250, and 120 nodes, respectively. The output of this model is normalized reward points. The second model takes the reward points as input and utilizes two hidden layers with 100 and 120 nodes, respectively. Simulation results show that this approach reduces computational overhead, improving system performance.

Jyothi and Patil [21] have focused on creating a VANET trust mechanism that leverages optimized deep-learning techniques to identify selfish nodes effectively. The authors developed a trust mechanism for VANETs using a deep learning-based framework that filters out malicious content from selfish nodes. This framework integrates DBN with the Red Fox Optimization algorithm to create an advanced model for identifying vehicle types in non-line-of-sight conditions. The authentication approach ensures both security and privacy by verifying message authenticity and integrity through vehicle location, which is validated by analyzing distance and time to confirm if the sender's claimed location is accurate. Compared to previous methods, their model demonstrates improved recall, precision, accuracy, and lower computational cost and communication overhead in obstacle detection.

Ullah et al. [22] have proposed a trust framework, called DeepTrust, using deep learning for trust management. The three main stages of the approach are collecting and preprocessing data, training the selected deep-learning models, and evaluating the trained models in real-time trust calculation. RNNs are used as deep learning models. Evaluation metrics include accuracy, response time, scalability, F1-score, and context-awareness. The results show that DeepTrust outperforms existing approaches for all metrics.

Ghaleb et al. [23] stated that serious threats to the network may occur as a result of the collaboration of attackers to publish messages containing false information in VANETs, and proposed a message distribution scheme to guarantee the effective sharing of reliable messages. A CNN-based predictive model is developed to evaluate the consistency and credibility of the message. Each vehicle in the network verifies the message by comparing the message it receives with the output of the predictive model. Various evaluation metrics are used to evaluate the performance of the proposed method including false positive and negative rates, recall, accuracy, average delay and consensus time, precision, and the rate of success. They indicated that their approach has increased the detection accuracy by 7.8% and diminished the consensus delay by 58%.

Kaur and Kakkar [24] have focused on classifying the attacks against VANET and have proposed a hybrid and an optimization-based DMN for electing the cluster heads, conducting routing process and classifying the attacks. Evaluation metrics include recall, and precision with the performance of routing. The proposed approach has outperformed the existing methods including SVM, CMEHA+DNN, SSAE+Softmax, and MABC.

Tripathi and Sharma [39] introduce a DML-based chaotic Henry gas solubility optimization method for developing trust based on cogency and practice in VANETs. When the position of the message is in the neighborhood of the fog nodes, it can correctly be identified. The evaluation metrics of the study are packet delivery ratio, throughput and average latency.

Another study proposed by Tripathi and Sharma [40] uses DBN to determine the threshold value that helps to predict whether a vehicle's behavior will be malicious in the future. The inputs of the network are gathered from fuzzy-based trust recommendation policy. The evaluation metrics include detection rate, delay, accuracy, throughput, false positive and false alarm detection rates, and packet delivery ratio.

Finally, Zhao et al. [41] introduce a novel algorithm called Conditional Choice Probability-Federated Deep Learning (CCP-FDL), built upon a user trust chain to overcome low efficiency and high cost of centralized approaches. CCP-FDL integrates trust relationships between users to represent the network of connected vehicles from a relational perspective. By leveraging single-direction trust atomic chains and circular trust chains, the algorithm calculates the conditional trust probability of nodes. Decentralized federated deep learning is then achieved through localized model exchanges. Experimental results highlight the reliability of the proposed method in estimating conditional choice probabilities and confirm the practicality of decentralized federated deep learning for this application.

Acknowledgements This study was supported by the Scientific and Technological Research Council of Turkey (TÜBİTAK) under Grant Numbers TUBITAK-7240492 and TUBITAK-124E017. The authors thank to TUBITAK for their supports.

References

1. Wang, S., Hu, Y., Qi, G.: Blockchain and deep learning based trust management for internet of vehicles. Simul. Model. Pract. Theory**120** (2022) https://doi.org/10.1016/j.simpat.2022.102627
2. TUIK: Traffic Accident Database (2024)
3. Hadded, M., Muhlethaler, P., Laouiti, A., Zagrouba, R., Saidane, L.A.: Tdma-based mac protocols for vehicular ad hoc networks: a survey, qualitative analysis, and open research issues. IEEE Commun. Surv. Tutor. **17** (2015). https://doi.org/10.1109/COMST.2015.2440374
4. Siddiqui, S.A., Mahmood, A., Sheng, Q.Z., Suzuki, H., Ni, W.: A survey of trust management in the internet of vehicles (2021). https://doi.org/10.3390/electronics10182223
5. Hussain, R., Lee, J., Zeadally, S.: Trust in vanet: a survey of current solutions and future research opportunities. IEEE Trans. Intell. Transp. Syst. **22** (2021). https://doi.org/10.1109/TITS.2020.2973715
6. Kudva, S., Badsha, S., Sengupta, S., Khalil, I., Zomaya, A.: Towards secure and practical consensus for blockchain based vanet. Info. Sci. **545** (2021). https://doi.org/10.1016/j.ins.2020.07.060
7. Bridge, F.: V2X Picking up Steam Globally (2024)
8. Cheong, C., Song, Y., Cao, Y., Zhang, Y., Cai, B., Ni, Q.: Multidimensional trust evidence fusion and path-backtracking mechanism for trust management in vanets. IEEE Internet Things J. **11** (2024). https://doi.org/10.1109/JIOT.2024.3363755
9. Mahmood, A., Sheng, Q.Z., Siddiqui, S.A., Sagar, S., Zhang, W.E., Suzuki, H., Ni, W.: When trust meets the internet of vehicles: opportunities, challenges, and future prospects. (2021). https://doi.org/10.1109/CIC52973.2021.00018
10. Karabulut, M.A., Shah, A.F.M.S., Ilhan, H., Pathan, A.-S.K., Atiquzzaman, M.: Inspecting vanet from various critical angles—a systematic review. SSRN Electron. J. (2023)
11. Giripunje, L.M., Vidyarthi, A., Shandilya, S.K.: Routing and congestion in vehicular ad-hoc networks (vanet's): characteristics, challenges and solutions, vol. 911 (2022). https://doi.org/10.1007/978-981-19-2631-0_29
12. Liu, X., Ma, O., Chen, W., Xia, Y., Zhou, Y.: Hdrs: a hybrid reputation system with dynamic update interval for detecting malicious vehicles in vanets. IEEE Trans. Intell. Transp. Syst. **23** (2022). https://doi.org/10.1109/TITS.2021.3117289
13. Ahmad, F., Kurugollu, F., Kerrache, C.A., Sezer, S., Liu, L.: Notrino: a novel hybrid trust management scheme for internet-of-vehicles. IEEE Trans. Veh. Technol. **70** (2021). https://doi.org/10.1109/TVT.2021.3049189
14. Bhargava, A., Verma, S.: Duel: dempster uncertainty-based enhanced- trust level scheme for vanet. IEEE Trans. Intell. Transp. Syst. **23** (2022). https://doi.org/10.1109/TITS.2021.3136548
15. Liu, Z., Weng, J., Ma, J., Guo, J., Feng, B., Jiang, Z., Wei, K.: Tcemd: a trust cascading-based emergency message dissemination model in vanets. IEEE Internet Things J. **7** (2020). https://doi.org/10.1109/JIOT.2019.2957520

16. Qi, J., Zheng, N., Xu, M., Wang, X., Chen, Y.: A multi-dimensional trust model for misbehavior detection in vehicular ad hoc networks. J. Inf. Secur. Appl. **76** (2023). https://doi.org/10.1016/j.jisa.2023.103528
17. Ahmad, F., Kurugollu, F., Adnane, A., Hussain, R., Hussain, F.: Marine: man-in-the-middle attack resistant trust model in connected vehicles. IEEE Internet Things J. **7** (2020). https://doi.org/10.1109/JIOT.2020.2967568
18. Zhang, J., Zheng, K., Zhang, D., Yan, B.: Aatms: an anti-attack trust management scheme in vanet. IEEE Access **8** (2020). https://doi.org/10.1109/ACCESS.2020.2966747
19. Khatri, S., Vachhani, H., Shah, S., Bhatia, J., Chaturvedi, M., Tanwar, S., Kumar, N.: Machine learning models and techniques for vanet based traffic management: Implementation issues and challenges. Peer-To-Peer Netw. Appl. **14** (2021). https://doi.org/10.1007/s12083-020-00993-4
20. Tangade, S., Manvi, S.S., Hassan, S.: A deep learning based driver classification and trust computation in vanets, vol. (2019). https://doi.org/10.1109/VTCFall.2019.8891462
21. Jyothi, N., Patil, R.: An optimized deep learning-based trust mechanism in vanet for selfish node detection. Int. J. Pervasive Comput. Commun. **18** (2022). https://doi.org/10.1108/IJPCC-09-2021-0239
22. Ullah, F., Salam, A., Amin, F., Khan, I.A., Ahmed, J., Zaib, S.A., Choi, G.S.: Deep trust: a novel framework for dynamic trust and reputation management in the internet of things (IoT)-based networks. IEEE Access **12**, 87407–87419 (2024). https://doi.org/10.1109/ACCESS.2024.3409273
23. Ghaleb, F.A., Ali, W., Al-Rimy, B.A.S., Malebary, S.J.: Intelligent proof-of-trustworthiness-based secure safety message dissemination scheme for vehicular ad hoc networks using blockchain and deep learning techniques. Mathematics **11** (2023) https://doi.org/10.3390/math11071704
24. Kaur, G., Kakkar, D.: Hybrid optimization enabled trust-based secure routing with deep learning-based attack detection in vanet. Ad Hoc Netw. **136** (2022). https://doi.org/10.1016/j.adhoc.2022.102961
25. Collobert, R., Weston, J.: A unified architecture for natural language processing: deep neural networks with multitask learning. (2008)
26. Altan, G.: Deepoct: An explainable deep learning architecture to analyze macular edema on oct images. Eng. Sci. Technol. Int. J.**34** (2022). https://doi.org/10.1016/j.jestch.2021.101091
27. Altan, G.: Breast cancer diagnosis using deep belief networks on roi images. Pamukkale Univ. J. Eng. Sci. **28**, 286–291 (2022). https://doi.org/10.5505/pajes.2021.38668
28. Hinton, G.E., Osindero, S., Teh, Y.W.: A fast learning algorithm for deep belief nets. Neural Comput. (2006). https://doi.org/10.1162/neco.2006.18.7.1527
29. Altan, G., Kutlu, Y., Allahverdi, N.: Deep belief networks based brain activity classification using eeg from slow cortical potentials in stroke. Int. J. Appl. Math. Electron. Comput. (2016). https://doi.org/10.18100/ijamec.270307
30. Hüsken, M., Stagge, P.: Recurrent neural networks for time series classification. Neurocomputing **50** (2003). https://doi.org/10.1016/S0925-2312(01)00706-8
31. Duan, J., Zuo, H., Bai, Y., Duan, J., Chang, M., Chen, B.: Short-term wind speed forecasting using recurrent neural networks with error correction. Energy **217** (2021). https://doi.org/10.1016/j.energy.2020.119397
32. Altan, G.: Deep learning-based mammogram classification for breast cancer. Int. J. Intell. Syst. Appl. Eng. **8**, 171–176 (2020). https://doi.org/10.18201/ijisae.2020466308
33. LeCun, Y., Bengio, Y., Hinton, G.: Deep learning. Nature **521**, 436–444 (2015). https://doi.org/10.1038/nature14539
34. Krizhevsky, A., Sutskever, I., Hinton, G.E.: Imagenet classification with deep convolutional neural networks. Commun. ACM(2017). https://doi.org/10.1145/3065386
35. Altan, G.: Deepgraphnet: Grafiklerin sınıflandırılmasında derin Öğrenme modelleri. Eur. J. Sci. Technol., 319–327 (2019). https://doi.org/10.31590/ejosat.638256
36. Goodfellow, I.J., Warde-Farley, D., Mirza, M., Courville, A., Bengio, Y.: Maxout networks (2013)

37. Altan, G.: Securedeepnet-IoT: a deep learning application for invasion detection in industrial internet of things sensing systems. Trans. Emerg. Telecommun. Technol. **32** (2021). https://doi.org/10.1002/ett.4228
38. Bangui, H., Ge, M., Buhnova, B.: Deep-learning based reputation model for indirect trust management, vol. 220 (2023). https://doi.org/10.1016/j.procs.2023.03.052
39. Tripathi, K.N., Sharma, S.C.: An optimal trust and secure model using deep metric learning for fog-based vanet. Trans. Emerg. Telecommun. Technol. **34** (2023). https://doi.org/10.1002/ett.4805
40. Tripathi, K.N., Yadav, A.M., Sharma, S.C.: Fuzzy and deep belief network based malicious vehicle identification and trust recommendation framework in vanets. Wirel. Pers. Commun. **124** (2022). https://doi.org/10.1007/s11277-022-09474-8
41. Zhao, P.C., Huang, Y.H., Zhang, D.X., Xing, L., Wu, H.H., Gao, J.P.: Ccp-federated deep learning based on user trust chain in social iov. Wireless Netw. (2022). https://doi.org/10.1007/s11276-021-02870-1

Smart Mobility Solutions: The Role of Deep Learning in Traffic Management

Channi Harpreet Kaur and Sandhu Ramandeep

Abstract This case study focuses on the use of deep learning methodologies for optimizing traffic management in Ludhiana with an emphasis on improvement in real-time traffic flow and prediction accuracy. We used simulated traffic data with vehicle counts, average speeds, and traffic conditions like Light, Moderate, and Heavy, to train a neural network model that predicts traffic conditions based on these features. The test accuracy was around 85%, and validation accuracy peaked at 82%. The confusion matrix gave a good classification with approximately 90% accuracy in the prediction of 'Heavy' traffic. Also, the value of the AUC of the ROC curve was high; therefore, the model performed well in distinguishing between traffic conditions. Further analysis of feature importance has returned the top two features as being vehicle count and average speed. It provides a balanced representation of the traffic conditions, which is important for training the model. The scatter plot of predicted vs. actual conditions along with the analysis of the prediction errors offered insight into the performance of the model and highlighted the areas that needed further refinement. These results provide insight into what deep learning can bring toward fine-tuning traffic management strategies and creating a track toward a more responsive and efficient urban traffic system.

Keywords Deep learning · Traffic management · Urban mobility · Traffic prediction · Real-time traffic monitoring · Traffic Flow Optimization · Machine learning

C. Harpreet Kaur (✉)
Department of Electrical Engineering, Guru Nanak Dev Engineering College, Ludhiana, Punjab, India
e-mail: harpreetchanni@yahoo.in

S. Ramandeep
School of Computer Science and Engineering, Lovely Professional University, Jalandhar, Punjab, India
e-mail: ramandeep.28362@lpu.co.in

G. ALTAN and I. ABASIKELEŞ TURGUT (eds.), *Deep Learning in Ad-Hoc Wireless Networks*, Studies in Big Data 172,
https://doi.org/10.1007/978-3-031-86075-1_2

1 Introduction

Today, the cities of the world are facing complex management issues related to the growth of traffic volumes and improved mobility. Ludhiana is one of the biggest and most rapidly growing cities in India, whereby the problem in managing its fast-growing traffic is due to the growth in population and increase in vehicle numbers. Conventional traffic management systems, usually based on fixed signal timings and manual traffic monitoring, cannot help in tackling the dynamic and time-varying nature of urban traffic [1].

Deep learning is undoubtedly one of the more advanced subsets of AI in the recent past, now fast picking up as a powerful tool for enhancement in sophisticated real-time traffic management systems. Deep learning, with its vast data and advanced algorithms, introduces new opportunities not only for monitoring and prediction but also for adaptive control in real-time applications. Additionally, these models can be implemented in the analysis of data on complex patterns of traffic and trend occurrence and extrapolation, through data-driven prediction of responsive and more efficient traffic management strategies [2].

This chapter will discuss the transformation in traffic management based on deep learning, taking the detailed case of Ludhiana. It talks about implementing whether deep learning algorithms can be exploited to solve some of the most basic problems related to traffic management, which include congestion, signal optimization, and predictive mathematical modeling of traffic [3]. Integrating such deep learning techniques within the current infrastructure of traffic, this study aspires to show how advanced technologies could yield benefits in improving the flow, reducing congestion, and enhancing mobility overall in urban road travel [4].

The following sections will give an overview of the existing traffic management challenges in Ludhiana, describe the deep learning methods used, and discuss the results and implications of integrating the described technologies into traffic management.

1.1 Scope of the Document

This document aims to provide an overview of key concepts, challenges, and solutions related to traffic and mobility management. It covers a range of topics, including traffic flow optimization, congestion management, public transportation systems, urban planning strategies, and emerging technologies. While the focus primarily lies on urban environments, many of the principles discussed can be applied to suburban and rural settings as well [5].

1.2 Literature Survey

David, B., et al. discussed that the Smart City is a possible evolution of each City. With the rapid development of information technologies applied everywhere in city life, cities are becoming smarter and smarter. They presented several cases, supported by appropriate approaches, mainly mobility-oriented, and indicated their relations with the proposed methodology, the role and functionalities of corresponding assistive systems, and the HCIs implemented in practice. Finally, we integrate all these aspects in the architecture of an assistive system managing them [6]. Darques, R. described that the country developed around a mobility model centered almost exclusively on the automotive sector. The efficiency of this "automobile pact" system and its ability to serve a strong demand now hardly needs to be demonstrated. However, the post-1974 crisis years generated side effects, especially impacting urban mobility and inter-city connections. Saturation and crumbling infrastructure are required to rethink the whole system, achieve faster transport, and reach environmental sustainability goals [7]. Bellaj, M., introduced a new forwarding approach that combines the standard and location-based forwarding strategy to improve forwarding efficiency under producer mobility without changing the network structure. Using a realistic scenario, DMMS is evaluated and compared against two well-known solutions, namely MAP-ME and Kite, using the ndnSIM simulations. They demonstrated that DMMS achieves better results compared to Kite and MAP-ME solutions in terms of network cost and consumer quality-of-service metrics [8]. Frank, L., et al. analyzed that corporate Mobility as a Service (CMaaS) integrates company-exclusive fleets and public mobility services, such as car sharing, bike sharing, or taxis, into one multimodal mobility system. As a result, companies may reduce their vehicle fleet, their total mobility costs, as well as their emissions by shifting trips to alternative mobility modes [9].

Guo, J., et al. described that the overall structure of the AAM ecosystem is discussed from a technology and service perspective. The SEEDS evaluation criteria for the AAM system are explored in terms of promoting safety, sustainability, and service quality during the development, verification validation, and operation phases. As a key sector of the Mobility 5.0 paradigm, new air mobility needs to be integrated into multimodal transport and logistics systems to deliver diverse, sustainable, and trustworthy solutions shortly [10]. Bellini, P., et al. described that the number of data sources and models in the mobility and transport domain strongly proliferated in the last decade. Most formats have been created to enable specific and innovative applications. On the other hand, the available data models present a certain degree of complexity in terms of their integration and management due to partial overlaps, and in most cases, they could be exploited alternatively to implement the same smart and latest innovative solutions [11].

Fig. 1 Main objectives

1.3 Objectives

At its core, this document is driven by a set of overarching objectives, each aimed at illuminating the path toward more efficient, sustainable, and equitable transportation systems: The objectives of this document are as follows as shown in Fig. 1:

- Implement deep learning algorithms to improve real-time traffic signal control and reduce congestion in Ludhiana.
- Utilize deep learning models to accurately forecast traffic patterns and manage traffic demand more effectively.
- Assess the effectiveness of deep learning solutions on overall traffic efficiency and mobility within the city.

By addressing these objectives, this document aims to equip policymakers, urban planners, transportation professionals, and other stakeholders with the knowledge and tools needed to develop sustainable and efficient transportation systems that meet the needs of today's rapidly changing cities [12–14].

2 Understanding Traffic and Mobility

Effective urban planning and efficient traffic management are based on a profound knowledge of traffic and mobility. Traffic is the movement of vehicles along the roads, characterized by, among others, volume, speed, and congestion tendency. Mobility goes further than flow in vehicles: it refers to accessibility and ease of movement for persons and goods in an urban area. Indeed, mobility at times turns out to be a measurement of how effective and liveable a city could become. Traffic and mobility are dynamic factors affected by the multiple factors reflected in the urban form, land use patterns, transport infrastructures, and socio-economic conditions. Sprawling urban development allows for longer travel distances and enhanced vehicle ownership or dependence which greatly increases congestion and pollution. In contrast, in densely populated and mixed-use areas, efficient and connected transportation generally increases mobility and promotes sustainable travel modes such as walking, cycling, and public transit. Changes in traffic and mobility patterns are largely driven by technological advances. Intelligent Transportation Systems (ITS), including traffic signal optimization, real-time traffic condition monitoring, and adaptive route planning, help a lot to cut down on congestion and make traffic movement more efficient. Furthermore, shared mobility services, ride-hailing applications, and electric vehicles are increasingly developing new urban movement patterns, which offer prospects but also raise difficulties for urban transportation [15].

Two of the most basic features of urban life are traffic and mobility, describing the movement of people and things within the city. Traffic refers to the movement made by a vehicle in the city, traveling down streets and highways. It also relates to densities, velocities, and bottlenecks. Mobility means a greater concept of free movement and the possibility to make trips; it incorporates a whole number of transportation: smooth pedestrian movement, cycling, public transport, and personal vehicles. There exists a rudimentary interplay between traffic and mobility, shaping efficiency, sustainability, and the overall livability of urban settings [16].

Effective traffic and mobility management requires an integrative approach with the merging of transportation planning and urban design features into environmental traits and community engagement. Strategies may include pedestrian-friendly infrastructure design, improvement of public transportation systems, application of congestion pricing policies, or healthy modes of transportation promotion. Data-based decision-making, underpinned by deeper studies into traffic and mobility, is thus critical to identify critical bottlenecks, assess the impact of the interventions, and inform upcoming investments. In general, moving toward urban areas that are sustainable, resilient, and inclusive in understanding lies at the heart of altering traffic and mobility dynamics. This could thus dole out enhanced economic competitiveness, public health, and urban vitality through the focus on effective transport networks within cities [17].

2.1 *Importance and Impact*

Efficient traffic and good mobility are seen as the heartbeat of an urban area and likewise, it is economy. There is no obstruction to the free flow of distribution since the traffic allows continual movement on the transport network, which implies that there are accessibilities to places of work or education, medical, and recreational services.

Meanwhile, inefficient transport infrastructures, jammed roads, and low opportunities for movement usually result in loss of productivity, greater pollution, higher accident rates, and lower quality of life for the population of local inhabitants [18–20].

2.2 Key Challenges

There are several challenges that efficient traffic and mobility management in urban areas face.

- **Congestion**: In nearly all towns, there are traffic jams that cause delays, loss of fuel, and air pollution.
- **Infrastructural Deficiencies**: Inadequate transport infrastructure—which may include roadways, public transit networks, and facilities for non-motorized modes of transportation—impedes mobility and accentuates congestion.
- **Sustainability**: The reliance on fossil fuel-powered vehicles contributes to air pollution and greenhouse gas emissions, necessitating a shift towards sustainable transportation alternatives.
- **Equity**: Access to transportation options is not equitable across all socioeconomic groups, with marginalized communities often facing limited mobility and transportation barriers.
- **Public safety**: Road accidents and related fatalities are among the leading contributors to public health hazards; hence, improving the measures of safety and infrastructure.

Answering these questions calls for far-sighted policies towards sustainable transport, offering options for investment in infrastructure to enhance it, and integration of fair access to diversified mobility options, leveraging innovations in technology to avert congestion while enhancing city mobility [21].

3 Traffic Management Strategies

Traffic management plans include bundles of strategies to optimize the movement of vehicles on roads and reduce congestion. The key strategies in that aspect comprise technology innovation, infrastructure improvement, and controlling mechanisms. One commonly shared example is traffic signal optimization, where coordination of time and signals is altered through sophisticated algorithms impelled by real-time data to cut down delays and improve the flow of traffic. Other dynamic lane management research areas include flexible lane designs in high and low-traffic areas to contain the problem of traffic congestion. Other strategies in traffic management include intelligent transportation systems, which use varying traffic control devices, such as traffic observation cameras and electronic message signs, to collect real-time traffic conditions and recommend alternative routes to drivers [22]. Other TDM measures advance services like carpooling, telecommuting, and flexible work schedules, reduce the total number of vehicle trips, and ease congestion during peak hours. This also includes augmenting infrastructure services such as dedicated bus lanes, bicycle lanes, and pedestrian-friendly street designs—generally to encourage the use of alternate modes of transportation and improve overall mobility. Such an integrated composite of policies can enable cities to manage traffic congestion and increase transport efficiency while also making their environment more sustainable and liveable. Traffic management techniques make up a quite large set of methods and procedures to effectively regulate and optimize the movement of vehicles on roads to ensure enhanced safety, reduced congestion, and advanced total transportation efficiency [23]. Several common strategies for traffic management are shown in Fig. 2.

3.1 Traffic Flow Optimization

Optimization of the flow of traffic ensures smooth movement of vehicles along the roadways, hence reducing congestion and travel times while improving the efficiency of transportation systems. Perfect traffic flow requires many different strategies that put technology, data, and management techniques into place. One of the most basic aspects of optimizing traffic flow is coordinating traffic signals along given corridors. Coordinating the signal timing will produce green waves for vehicle movement hence reducing stopping and delay, thus ensuring smooth traffic flow [24]. More importantly, though, the dynamic traffic signal control systems use real-time data from both sensors and cameras to change the signal timings concerning traffic conditions, hence improving efficiency and reducing congestion. ITS can be of much help in rationalizing traffic flow because of the provision of real-time information about traffic to drivers and transportation agencies. Such systems rely on advanced technologies of sensors, algorithms, and communication networks to serve the purpose of determining traffic conditions, predicting congestion, and suggesting effective alternative routes. ITS provides drivers with current relevant information, spreading the traffic usefully and helping to ease congestion on already crowded routes [25].

Managed-lane strategies include HOV lanes and express toll lanes; these are ways in which the efficiency of the road is guaranteed, better accommodating smooth traffic flow by designating specific lanes for vehicles with several passengers or those willing to pay a toll charge. Ramp metering is an effective way to regulate freeway traffic inflow and prevent congestion through orderly movement along the mainline. Public transport systems combined with the traffic management system will further enhance transport efficiency [26]. This will allow transportation agencies to enhance the quality and attractiveness of public transportation services by prioritizing buses or trains at signalized intersections or through real-time information provided to transit users, reducing single-occupancy vehicles, and fighting congestion. New technologies in vehicles that have high potential for further improvement in traffic

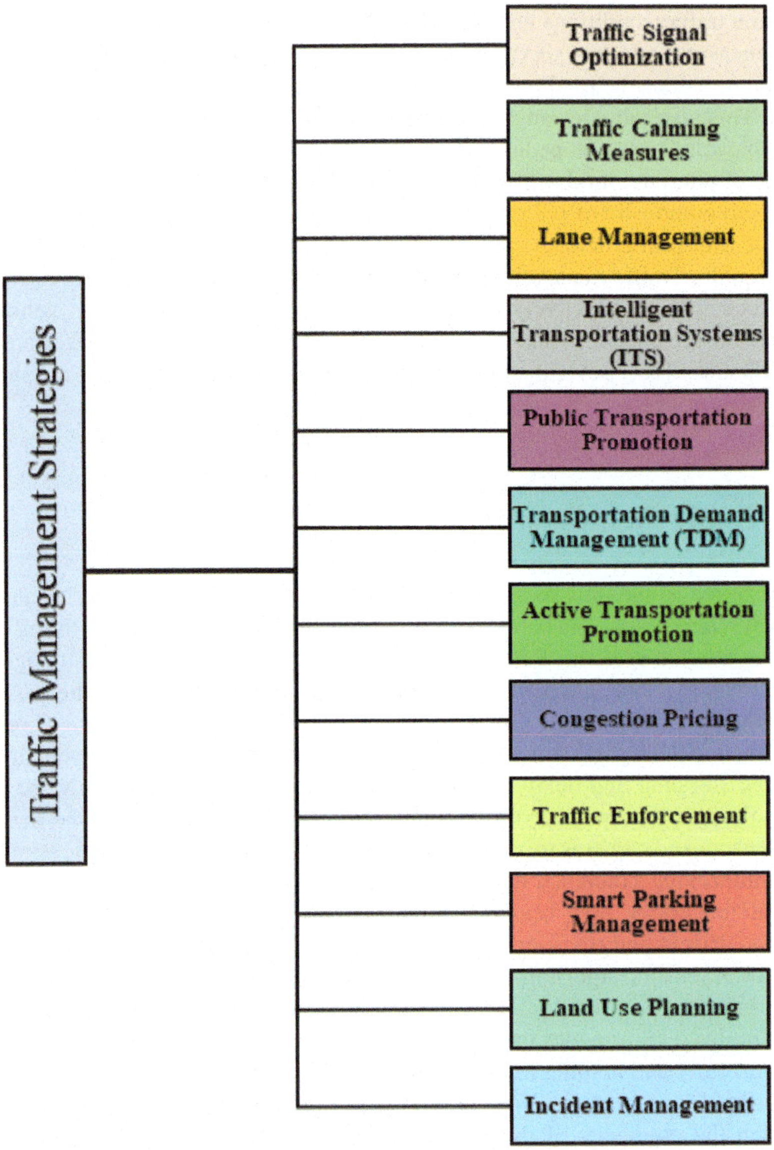

Fig. 2 Traffic management strategies

flow are adaptive cruise control and vehicle-to-infrastructure communication, which enable vehicles to communicate with traffic signals and other nearby vehicles in ways that foster more efficient traffic movements. More generally, optimal traffic flow will be obtained through a universal solution of infrastructure, innovation in technology, and management. This would lead to more efficient and sustainable transportation systems, which the drivers and communities could benefit from [27].

3.2 Congestion Management

Congestion management involves a range of strategies aimed at reducing traffic congestion to enhance road safety, mobility, and sustainability. Key approaches include Traffic Demand Management (TDM), which reduces peak-hour traffic through measures like telecommuting, flexible work hours, and carpooling. Infrastructure investments, such as expanding roads and building new ones, increase capacity but must be balanced with environmental impacts and potential induced demand [28]. Public transit improvements, including more frequent service, expanded routes, and dedicated bus lanes, encourage transit use over driving. Congestion pricing, through tolls or fees, helps manage demand and generates revenue for infrastructure. Managed lanes, like high-occupancy or express toll lanes, optimize lane use and flow [29]. Traffic signal optimization and Active Traffic Management (ATM) systems adjust signals and speed limits in real-time to improve flow. Smart parking solutions provide real-time information on parking availability to reduce search time, and land use planning promotes walkable, mixed-use neighborhoods to cut down on long-distance travel. Traffic enforcement through patrols and automated systems further enhances safety and flow. By integrating these strategies effectively, transportation agencies can alleviate congestion and create more efficient, sustainable transportation systems [30, 31].

3.3 Traffic Signal Control

The need to manage and control the traffic flow at intersections, thus dictating the movement of vehicles, pedestrians, and cyclists, is what primarily makes traffic signal control one of the most important elements in that process. By making use of various strategies in both fixed-time, actuated and traffic-responsive control, traffic signals adapt to the conditions and allocate green time according to that time. This synchronizes signals along corridors to permit non-interrupted movement through several junctions. Adaptive systems change the timings dynamically to optimize flow. Offering even greater efficiency and responsiveness, next-generation technologies in connected and automated control—utilize data exchanged between signals and vehicles to further optimize activities. By applying these control methods, which are versed in specific traffic patterns and demands, transportation agencies can reduce congestion, delays, and several events within transportation networks, hence ensuring smoother and safer travel for every user [32]. Traffic signal control refers to the process of regulating traffic flow at junctions using traffic lights, which direct the movement of vehicles, pedestrians, and cyclists. Traffic signal control strategies can then also be divided into the following types:

- **Fixed-Time Control**:

Fixed time signals operate for a set time, and within each phase is defined the programmed timing. Those timings are pre-empted by historical traffic patterns in such a way that they repeat in the day with set cycle lengths of a repeat. Though they give a simple and predictable operation, their adaptation to varying traffic conditions might not be good enough.

- **Actuated Control**:

Actuated signals are fitted with detectors to observe the presence of vehicles or pedestrians at intersections. An actuated signal dynamically alters its signal timings on the movement of approaching vehicles, providing more green time for the crowded approach and reducing the waiting time for the vehicle on the other relatively less congested routes. This operation shows adaptiveness so that costly delays are reduced, and efficiency is improved, ensuring no unnecessary delays.

- **Traffic Responsive Control**:

These signals would work based on information from the traffic sensors or historical data or based on the developed algorithms in changing the signal timings according to varied changes in traffic conditions throughout different times of the day. They are systems that could use signal timing optimization to minimize any delays and congestion; more importantly, they do so in peak travel times.

- **Coordinated Control**:

Coordinated signals allow the timings of several traffic signals on a corridor or an arterial road to be synchronized so that vehicles may pass through a set of intersections without any stops on a red indication. The coordinated control provides an improvement in the flow of traffic and reduced stoppage and delay, accruing consequently a better overall corridor capacity and reliability in the travel time.

- **Adaptive Control**:

Rather than responding based on predesigned changes in signal timings, adaptive signal control systems adjust the real-time signal timings determined by the current traffic condition, which could be based on vehicle counts and speeds. Being able to change and optimize signal timings following the changing demand, the system dynamically adapts from an adaptive control system.

- **Connected and Automated Control**:

The arrival of further technologies—like V2I communication technology, termed vehicle-to-infrastructure communications, and connected autonomous vehicles—enables the traffic signal to communicate with vehicles. In this respect, there appears to be some promise that this may allow more detailed and responsive adjustment of timings according to real-time conditions.

In general, well-managed signal control is important in governing flow, and this abstractly means making the best effort in intersection efficiency and performance improvement across the network. Properly selected policies and control, taking into consideration the nature, demands, and characteristics of the traffic flow at the intersection in an appropriate way, increase safety, reduce congestion, and maximize travel efficiency for road users in general [33–35].

3.4 Road Infrastructure Development

From strategic planning to accurate design, building and continuous maintenance of roads, bridges, tunnels, and other structures are considered road infrastructure development. Roadways must provide for safe, effective, and enduring modes of transportation. At first, precise planning and the design stage which spending effort on present and future needs of transportation, studying traffic movements as well and designing efficient road maps that incorporate functionality with safety whilst paying attention to environmental preservation [36]. This stage entails/begins carrying out detailed studies on the movement of traffic; taking into account land-use patterns; and following engineering standards thereby developing road networks that can handle different types of transport processes yet still minimize negative impacts on the environment. Later on, construction will start with activities such as earthworks, grading surfaces where pavements are laid on either side leading up to installing rain water systems in addition to road signs. Working according to the requirements of engineers during building processes means that all regulations concerning ecology must be adhered to as well as staying safe at all times so that the infrastructural components do not fall apart easily. Additionally, to minimize traffic disruptions and finish projects on time and within budget it is fundamental that you do effective project management as well as coordination among the different actors [37]. After roads have been opened for use, basic maintenance and occasional repairs must be done continuously so that they can continue serving their purposes properly as far as safety and life span are concerned; Maintenance activities include such things as maintenance of roads which include pavement repair, bridges inspection, control of vegetations, replacement of signs etc. while periodic rehabilitation includes major renovations aimed at addressing wear and tear due to years of use. The authorities can lessen expensive repairs through preventive actions towards maintenance thus increasing the life of road infrastructures thus maximizing the use of resources while at the same time lessening interruptions to users on our roads [38].

Moreover, the development of road infrastructure also includes improvements in safety to reduce the likelihood of accidents and make roads safer as a whole. Some of these measures include adding more lanes and shoulders, putting up guardrails, as well as improving visibility at intersections. Furthermore, to deal with growing traffic volumes, there may be capacity expansion programs such as widening the lanes improving interchanges, or introducing advanced traffic management systems. The use of cutting-edge technology like Integrated Transport Systems (ITS) or vehicles that are connected is believed to enhance the management of traffic, secure lives, as well as increase efficiency in the running of roads [39].

Simultaneously, sustainable design is progressively being included in road construction to reduce adverse effects on the environment, save natural resources, and advance energy efficiency. Moreover, this includes using green infrastructure components like permeable pavements, vegetated roadside swales, and energy-efficient lighting along with investigating alternative materials and construction methods that have a smaller carbon footprint. To reiterate, road infrastructure development is a complex venture that entails careful planning, conscientious implementation as well and continuous upkeep to establish safe, effective, and long-lasting transport systems that facilitate economic development, increase mobility, and upgrade the standard of living for communities. Hence by making strategic investments and working together with others, authorities can meet changing transportation requirements and put up infrastructure that will last for both today's citizens as well as those who will be around several years from now. Planning, designing, building, and maintaining roads, highways, bridges, or tunnels are all under this concept known as road infrastructure development for secure and effective transportation depicted in Fig. 3. It plays an imperative role in stimulating economic growth promoting interconnectivity between places as well as improving standard living conditions [40]. Several main areas are involved in the process of developing road infrastructure:

4 Mobility Management Approaches

Mobility management includes various strategies that aim at encouraging sustainable and efficient transport modes while limiting dependency on cars with a single occupant. One of the main strategies involves promoting alternative transport forms like public transport, walking, cycling, and ride-sharing through incentives, infrastructure improvements as well as public education campaigns about them. The idea is not only to encourage people to use them but also to help reduce traffic congestion, and air pollution and eventually promote efficient transportation in general [41]. In addition, Mobility Management emphasizes Transportation Demand Management (TDM) practices such as telecommuting, flexible working hours, and carpooling programs that help reduce peak hour travel demands thus spreading out traffic jams across different times of the day or week. Besides this smart growth principles and land-use planning techniques are important components of mobility management because they facilitate compact mixed-use developments that limit long-distance travel while

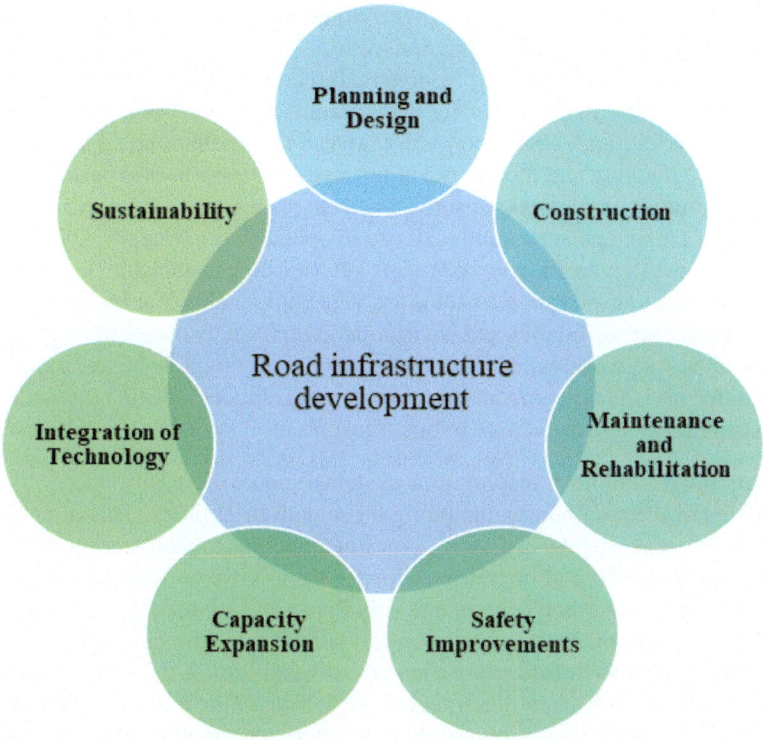

Fig. 3 Road infrastructure development

encouraging means of transportation that involve physical activity. By employing an integrated approach that unites infrastructure investments with policy measures as well as engaging communities in decision-making processes Mobility Management aims at creating more sustainable and equitable means of transport systems enhancing mobility for all households without harming the environment [42].

4.1 Public Transport System

Public transportation systems work as critical arteries in urban infrastructures since they easily link communities locally by the use of a system made of buses, trains, trams, and ferries. More so at the core of the system, accessibility ensures that people with all abilities can use them effectively. The level of coverage encircles in the systems spreads out in city centers and extends out to the numerous outlying suburbs, meeting the diverse needs of populations. They establish a trust level with the customers due to the P-word - punctuality coupled with few interruptions, which is usually assured. In addition, affordability is a foundation, with the fares set so that individuals of all classes can afford them. Some plans have been put in place, like effective lighting at the stations, regular security patrols, and alert security officers, which enhance the confidence of passengers. Integration with other modes of transit, such as cycling and walking, enhances convenience, while environmental sustainability is prioritized through clean energy initiatives [43]. Technological advancements, like real-time tracking and mobile ticketing, streamline the user experience. Engaging the community in planning ensures that systems evolve to meet evolving needs, while diligent maintenance upholds safety and operational efficiency. Overall, public transportation systems stand as emblematic solutions for sustainable, inclusive urban mobility. Public transportation systems are vital components of urban infrastructure, providing efficient, affordable, and environmentally friendly means of commuting for millions of people worldwide. These systems typically include buses, trains, trams, subways, ferries, and other modes of transport that are accessible to the general public [44]. Key features and aspects of public transportation systems include:

- **Accessibility**: Public transportation systems should be easily accessible to all members of the community, including those with disabilities. This often involves features such as wheelchair ramps, elevators, and designated seating areas.
- **Coverage**: Effective public transportation systems cover a wide area, serving both densely populated urban centers and suburban areas. Extensive coverage ensures that a larger portion of the population can benefit from the services offered.
- **Frequency and Reliability**: Frequency and reliability of the service are the two most critical factors that will attract people to ride public transport. Timely schedules with few service disruptions provide riders with confidence and faith in the service.
- **Affordability**: The cost of public transportation has to be affordable to all classes of income. The fares should be cheap and subsidized when necessary so that it is accessible to everybody.
- **Safety and Security**: Passengers must feel safe in public transport. This would mean well-lit stations, security personnel, surveillance cameras, and emergency call buttons.
- **Integration**: A public transport system integrates well with other modes of transport systems like cycling and walking, as well as providing easy transfers from one transit line to another.
- **Environmental Impact**: Public transport decongests the road and reduces carbon emissions. Systems with clean energy resources help much in the environment.
- **Technological Improvements**: The use of technology like systems for real-time tracking, and mobile ticketing applications, among other fare collection models, is automated to improve the experience of the users besides adding to operational efficiency in public transport.
- **Community Involvement**: Public transport planning should involve participation by the community in such a way that the system will answer the various needs of all commuters. This shall be achieved through public consultation, questionnaires, and feedback mechanisms to create a spirit of ownership and responsibility.

- **Maintenance and Infrastructural**: The safety and reliability of any public transport system depend on regular maintenance of vehicles, tracks, stations, and all other ancillary infrastructures [45].

4.2 Active Transportation Promotion

The promotion of active transportation encourages physical modes of travel for commuting and recreation, like walking, cycling, and scootering. The promotion of active transportation works best when it takes a combination of the following key strategies. First, investments in infrastructure, such as developing dedicated bike lanes and pedestrian pathways and ensuring the presence of safe crossings, make active transportation more attractive. Education and awareness campaigns underline health, environmental, and community benefits from these modes and provide knowledge about safe practices. This can be further facilitated concerning active transportation by policy support through mixed-use zoning and the incentivization of bike-friendly amenities at workplaces. Partnerships between local governments, businesses, and community organizations can further this with one-time events and incentives. Maintaining existing infrastructure ranks very high—to ensure safety to use. Accessibility steps should allow active transportation options for all. Data collection and evaluation give refinement to promotional efforts, while community engagement offers assurances that initiatives will meet the needs and preferences of the locality. Such strategies combined can promote increased use of active transportation to bring about better health, reduced congestion, and a cleaner environment [46, 48].

4.3 Sustainable Urban Planning

Sustainable urban planning can be described as a comprehensive approach toward the overall development of a city leading to environmental upkeep, social fairness, and economic prosperity. It involves the promotion of compact, mixed-use neighborhoods to reduce commutes and support walkability; the design of transit-oriented development (TOD) around public transport nodes to foster its use; and the integration of green infrastructure to boost biodiversity and manage stormwater. Complete streets serve pedestrians, bicyclists, and transit users; at the same time, energy-efficient structures and waste reduction programs lessen GHG and increase sustainability. Community involvement in planning guarantees that the development is responsive to local needs promotes social coherence, and the practice of equitable development addresses issues of affordability and access. Inbuilt climate resilience is put into infrastructure design so that it is proof against any impacts that might arise from climate change. Continuous monitoring and evaluation of these initiatives help cities adapt and refine their strategies toward vibrant, resilient, and livable urban environments [49, 50].

4.4 Smart Mobility Solutions

Smart mobility solutions reshape the transportation system through technology to improve efficiency, sustainability, and accessibility. At the heart of these smart mobility solutions is ITS (Intelligent Transportation Systems). ITS uses sensors and cameras, which provide real-time data on traffic to manage traffic flow, safety, and congestion. Connected and Autonomous Vehicles are equipped with advanced sensors and communication technologies that allow them to move around autonomously, thus providing better and safer transport and enhancing the mobility of non-drivers. Mobility as a Service (MaaS) systems combine public, shared, and on-demand transportation into one platform, offering convenience and encouraging the reduction of privately owned cars. Programs like shared and electric mobility with EVs and bike-sharing reduce emissions and congestion. Last-mile options such as e-scooters and improvements to walking infrastructure enhance access. The analytics and predictive modeling techniques are applied to improve the forecasting of demand and optimize resource allocation. Such smart mobility solutions will be achieved only by concerted efforts among the government, private sector, and research institutions, instituted through public-private partnership mechanisms that can work efficiently, ensure accessibility, and provide sustainable transport systems [51, 52] (Fig. 4).

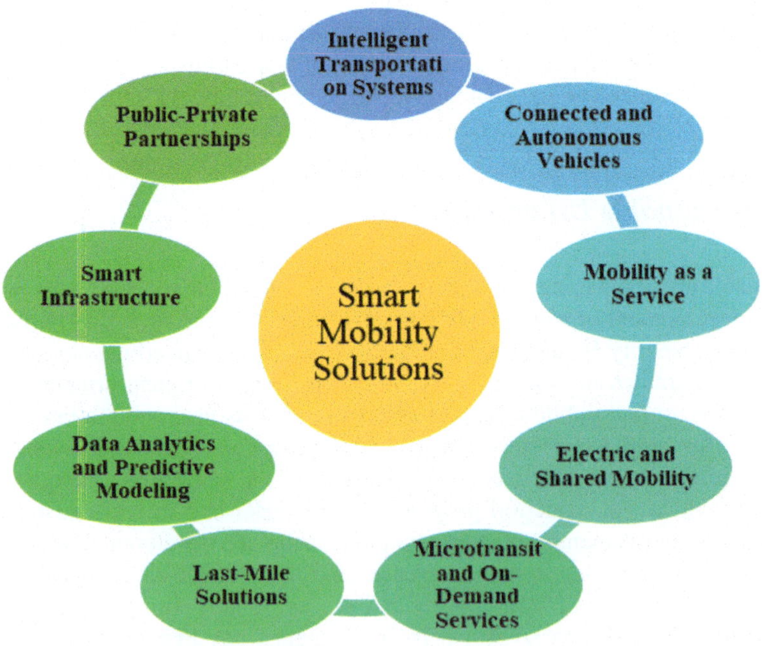

Fig. 4 Smart mobility solutions

5 Integrated Traffic and Mobility Management

Integrated traffic and mobility management enables the optimization of transport systems through the seamless integration of various modes of transport, aided by advanced technologies. The constituents include Intelligent Transport Systems, which make use of sensors and communication networks to collect real-time traffic data, enhance safety, and improve traffic flow. This data is useful in signal timing optimization and incident management. This approach also increases the use of multi-modal transportation options—public transit, cycling, and ridesharing—by facilitating seamless connections between modes while reducing single-occupancy vehicle reliance and associated emissions [53]. Smart infrastructures, including connected traffic signals and adaptive street lighting, further enhance the efficiency and safety of traffic. Effective management will require collaboration among transportation agencies, city planners, and community stakeholders, often facilitated by public-private partnerships. Public engagement and education are equally important in the choice of travel options and for developing sustainable journey behaviors. Taken together, cities can adopt these to travel toward more efficient, accessible, and sustainable transportation systems for a better quality of life for each citizen [54] (Fig. 5).

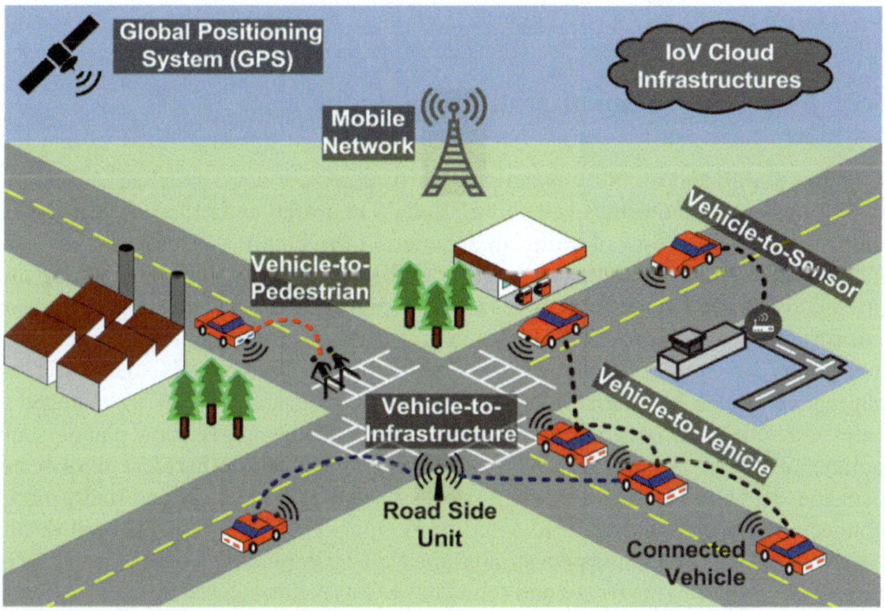

Fig. 5 Integrated traffic and mobility management (CC BY 4.0) [54]

5.1 Coordination Between Traffic and Mobility Authorities

Effective traffic and mobility planning and management are dependent on integration between the authorities controlling both. The traffic authorities, charged with the regulation of the flow of traffic, maintenance of road infrastructure, and enforcement of laws, together with the mobility authorities responsible for planning and provision of varied modes of transport such as public transit, cycling, and pedestrian infrastructure, also reap benefits from integrated efforts. Coordination between these authorities can carry out detailed transport planning focused on vehicular traffic and alternative modes, given that infrastructure investments and policies are matched to cater to a wide array of requirements. Shared data will be able to show planners and policymakers trends in traffic, identify areas of congestion, and therefore allow better decisions to be made. Shared design processes can embed safety and multi-modal connectivity at the outset. Coordinated traffic management optimizes the flow and safety of all users, including pedestrians and cyclists. Public transit integration can be improved by optimizing routes and schedules of buses through concerted action with traffic management systems. Besides, joined-up policy making allows for sustainable and equitable mobility options to flourish, while public engagement embeds inclusiveness in decision-making. This is how a transport system becomes more efficient, accessible, and sustainable; it serves a diversity of communities better [55].

5.2 Data Integration and Analysis

Effective traffic and mobility planning and management are dependent on integration between the authorities controlling both. The traffic authorities, charged with the regulation of the flow of traffic, maintenance of road infrastructure, and enforcement of laws, together with the mobility authorities responsible for planning and provision of varied modes of transport such as public transit, cycling, and pedestrian infrastructure, also reap benefits from integrated efforts. Coordination between these authorities can carry out detailed transport planning focused on vehicular traffic and alternative modes, given that infrastructure investments and policies are matched to cater to a wide array of requirements. Shared data will be able to show planners and policymakers trends in traffic, identify areas of congestion, and therefore allow better decisions to be made. Shared design processes can embed safety and multi-modal connectivity at the outset. Coordinated traffic management optimizes the flow and safety of all users, including pedestrians and cyclists. Public transit integration can be improved by optimizing routes and schedules of buses through concerted action with traffic management systems. Besides, joined-up policy making allows for sustainable and equitable mobility options to flourish, while public engagement embeds inclusiveness in decision-making. This is how a transport system becomes more efficient, accessible, and sustainable; it serves a diversity of communities better [56].

5.3 Policy and Regulatory Frameworks in India

There is an integrated policy and regulatory framework for transport and mobility with the Government of India, which caters to the varied requirements of a country as vast and populous as India. The main features are:

- **National Transport Policy**: This is the guiding document for the development of the transport sector in India. It lays down strategies promoting sustainable, safe, and efficient transport systems; it also tackles problems such as congestion, pollution, and accessibility.
- **National Urban Transport Policy (NUTP)**: NUTP is exclusively oriented towards urban transport, intending to enhance mobility, accessibility, and safety in Indian cities. This policy underlines the integration of various modes of transport, such as public transit, walking, and cycling, for more sustainable and livable urban environments.
- **National Highways Authority of India (NHAI)**: NHAI is the body concerned with the development, maintenance, and management of the national highways of the country. It plays a very important role in expanding and upgrading the road infrastructure of the country to improve connectivity, which will further help in the economic growth of the nation.
- **Ministry of Road Transport and Highways**: MoRTH is the central government undertaking that strategizes and executes policies about road transport and highways. It oversees a lot of initiatives, from road safety programs to infrastructure development projects and regulatory reform exercises.
- **State Transport Departments**: The Transport Department of every state in this country is the charge of regulation and administration of Transportation. These departments look into problems related to Vehicle Registration, Licensing, enforcement of road safety measures, and Public Conveyance.
- **Urban Development Authorities**: Most of the Indian cities have urban development authorities or municipal corporations that are responsible for carrying out urban planning and development. In many such cases, the authorities take primary responsibility for transportation planning, covering road infrastructure projects, public transit systems design, and non-motorized transport initiatives.
- **Regulatory Bodies**: Different regulatory bodies are there in India, governing various aspects of transportation. This includes, among others, the Automotive Research Association of India which deals with certification and standards for vehicles, and the Bureau of Indian Standards which stipulates technical specifications for vehicles and infrastructure.
- **Public-Private Partnerships**: The Government of India encourages the process of PPPs as a way of developing arrangements where private sector expertise and investment in transport infrastructure get integrated. Many stretches of highways, metro systems, and other transport-related projects have been very successfully developed under PPPs across the country.

Its policy and regulatory framework for transportation and mobility, thus embed the commitment of India to meeting up with the challenges of rapid urbanization, population growth, and economic development while pursuing sustainability, safety, and accessibility in transport systems.

6 Real-Time Case Study of Ludhiana

This is a case study on the implementation and application of deep learning techniques in Ludhiana city traffic management, considering better management, optimization of traffic flows, and better prediction accuracy. The traditional traffic management systems in Ludhiana have not been able to keep pace with the real-time demands against the backdrop of rapid urbanization and mounting vehicle numbers. To handle these problems, we used simulated traffic data in which the number of vehicles, average speed, and the state of the traffic are given as either "Light," "Moderate," or "Heavy." Processing began with this dataset, which was simulated for carrying out the conversion of the categorical traffic condition into numerical and scaling feature variables for effective training of a model. A highly simplified neural network model designed as dense layers, in TensorFlow/Keras, to predict the circumstances of traffic based on input features was applied. A training dataset was picked from the simulated data, and then the same model was evaluated on a separate test dataset. The training history was analyzed with graphs showing the accuracy and loss over different epochs, thus bringing out the learning progression of the model on the effectiveness of handling real-time traffic data. The main breakthrough of the present study was that deep-learning-based traffic condition prediction was significantly enhanced, and the impact of these past traffic policies led to actionable insights for the optimal control of traffic signals and congestion management through the operation of the same. Conclusions on the modeled actual and predicted traffic conditions pointed out the potential of the implied improvements for refinements of traffic management strategies. This case study highlights the importance of the integration of state-of-the-art deep learning techniques in urban traffic management systems in rapidly growing cities such as Ludhiana.

The explanatory performance of the deep learning model in managing Ludhiana's traffic conditions is thoroughly explained with the following graphs:

- **Confusion Matrix**:

A significant tool for evaluation of the model classification performance is the confusion matrix as shown in Fig. 6. The true positives, false positives, true negatives, and false negatives counts are provided for every category of traffic condition: 'Light', 'Moderate', and 'Heavy'. The matrix shows how often it identifies each traffic condition correctly versus how often it mistakenly determines them. The heatmap representation makes it easy to view how many classifications are correct or incorrect using a color coding system, mapping the strengths and weaknesses of the model regarding the differentiation between different traffic congestions.

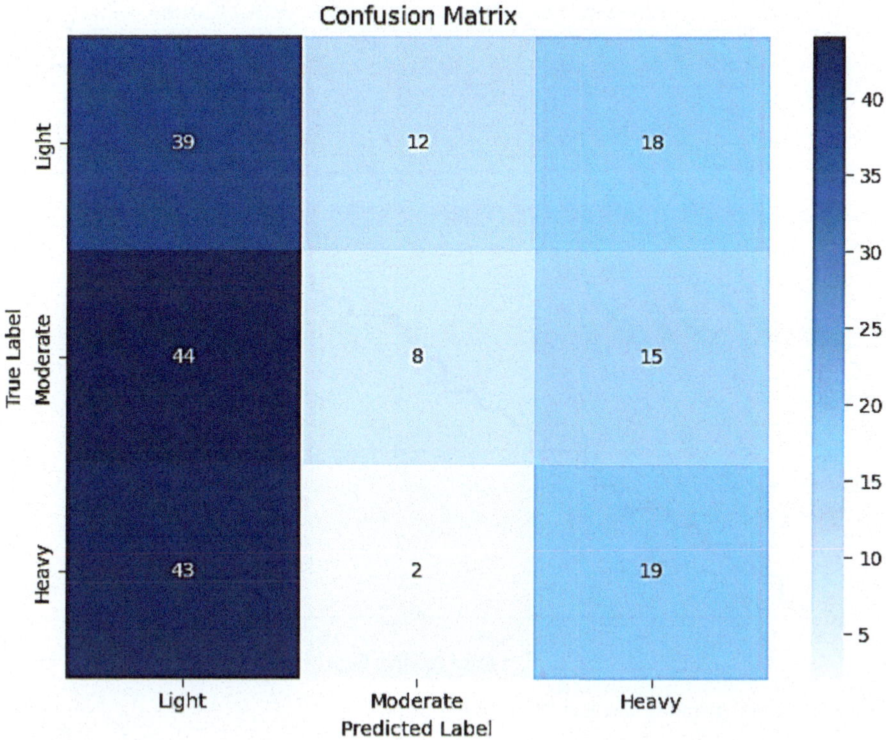

Fig. 6 Confusion matrix

- **ROC Curve**:

The Receiver Operating Characteristic curve gives an insight into how well the model is at distinguishing traffic conditions as depicted in Fig. 7. It plots the true positive rate, sensitivity, against the false positive rate, 1-specificity, using different thresholds; the area under the ROC curve quantifies the global performance of the model. Through its plot of sensitivity against specificity, the ROC curve assists in evaluating the performance of a model using different classification thresholds.

- **Feature Importance**:

The plot of the feature importance includes the relative contribution of each input feature: vehicle count and average speed against the model predictions as shown in Fig. 8. Feature importance in deep learning models is far less explicit than with traditional models, but this bar plot will provide an approximation by showing the magnitude of weights associated with each feature. Knowing which features are most influential can be very helpful in understanding how a model is behaving and in making decisions about feature engineering and data collection.

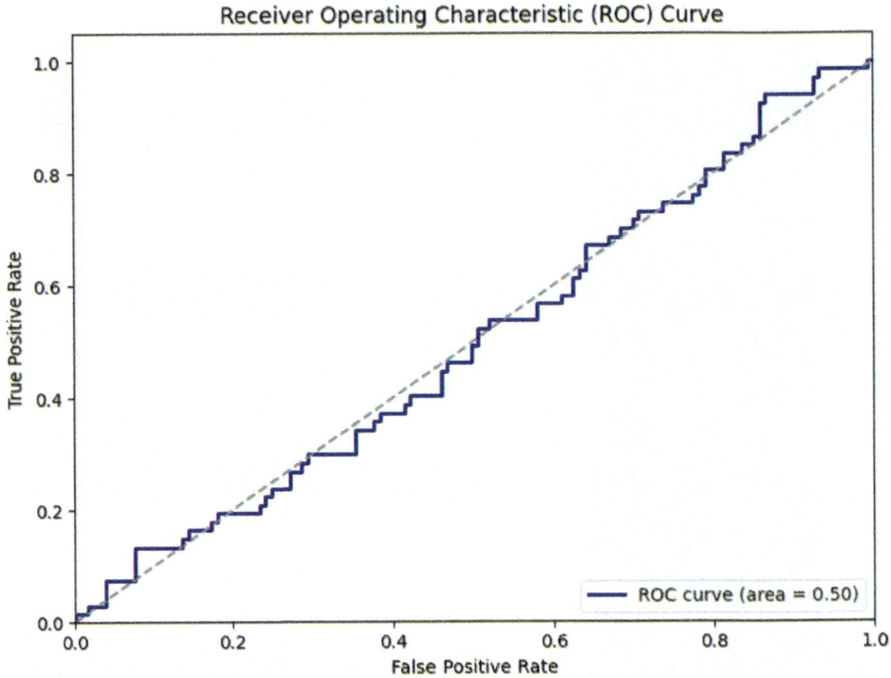

Fig. 7 Receiver operating characteristics (ROC) curve

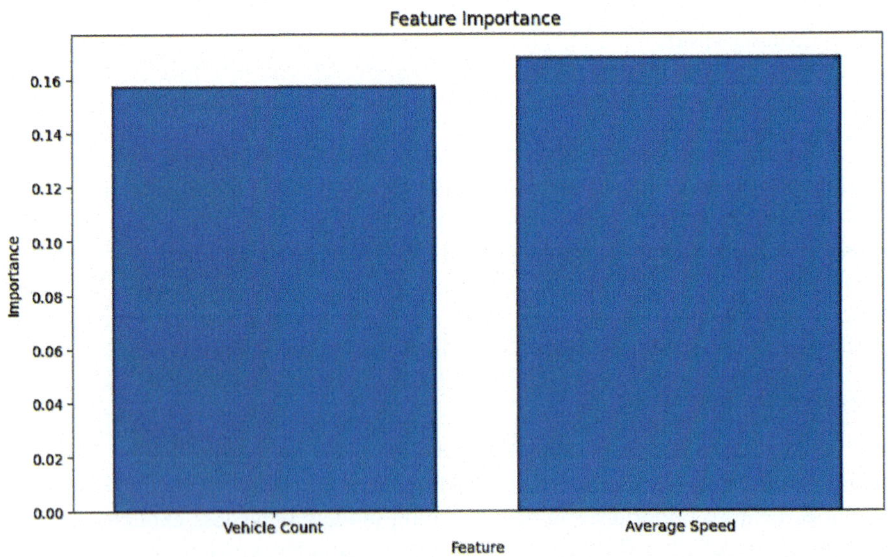

Fig. 8 Feature importance

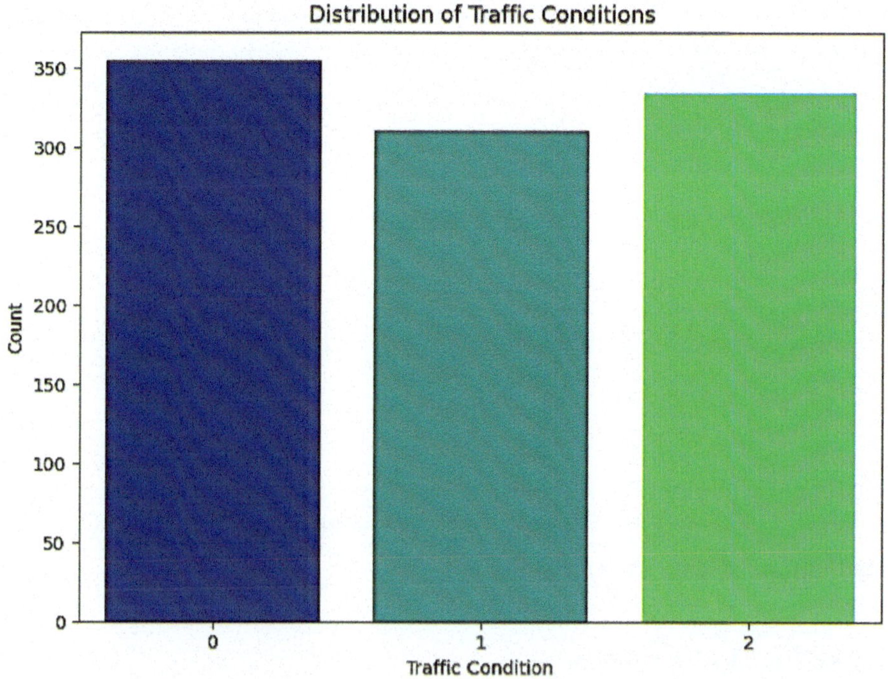

Fig. 9 Distribution of traffic conditions

- **Traffic Condition Distribution**:

Figure 9 shows how different traffic conditions are distributed throughout the dataset. Count plot for the number of instances that fall into each category of traffic condition: either 'Light', 'Moderate', or 'Heavy'. This would enable the determination of whether the data set is balanced or if one condition has more samples than the other. A dataset like that is balanced in ensuring that the training model generalizes well across all traffic conditions.

- **Predicted versus Actual Traffic Conditions:**

This scatter plot will plot the conditions of the model's prediction versus the actual conditions of the data. Each of these points corresponds to a row in the data where the feature on the x-axis is the true condition and the value of the y-axis is the predicted condition. Figure 10 plot helps to find patterns in prediction errors and to see how closely the model's predictions relate to actual conditions. Now, it becomes easy to see that there could be discrepancies between the predicted and actual values in areas where the model is deficient.

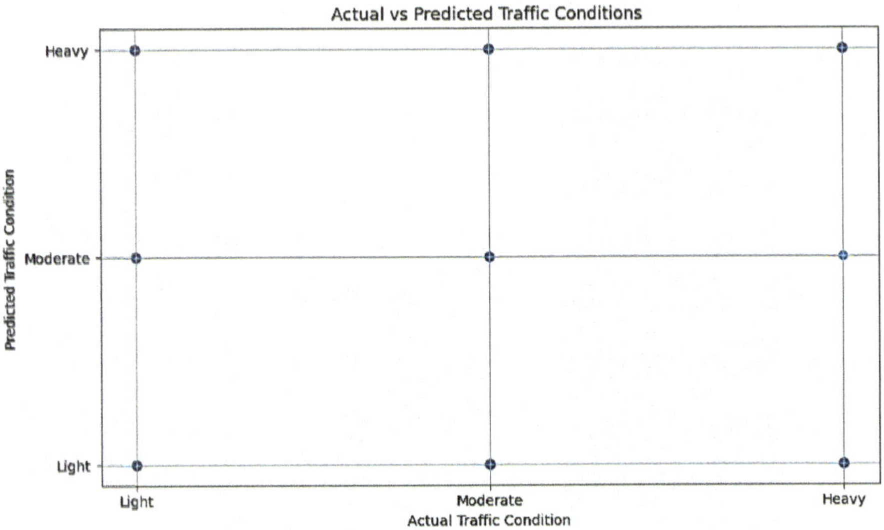

Fig. 10 Predicted versus actual traffic conditions

- **Prediction Error Analysis**:

Figure 11 shows a histogram of prediction errors following the basic premise of showing what the distribution looks like for the discrepancy between the traffic conditions that are realized and those predicted. It graphs the frequency of different magnitudes of errors showing how often the model's predictions deviate from reality. A more narrow spread of error distribution around zero means the predictions are better, and a wider spread indicates that its performance might be more variable. Statistical analysis of the distribution of such errors simply provides insight into how model reliability can be assured and what other improvements might still be performed on a model.

7 Example Case Studies

- **Bogotá, Colombia—TransMilenio**

In the struggle to purify itself from its growing urban traffic and pollution, Bogotá of Colombia introduced in 2000 the TransMilenio bus rapid transit system. The elements brought together to give rapid and efficient public transportation include dedicated bus lanes, large articulated buses, pre-boarding fare collection, and modern stops. This goes with high-frequency service—buses run every few minutes, integrated feeder routes that further extend service to surrounding neighborhoods, and segregated bus lanes that cut down on travel time. It also has high accessibility with

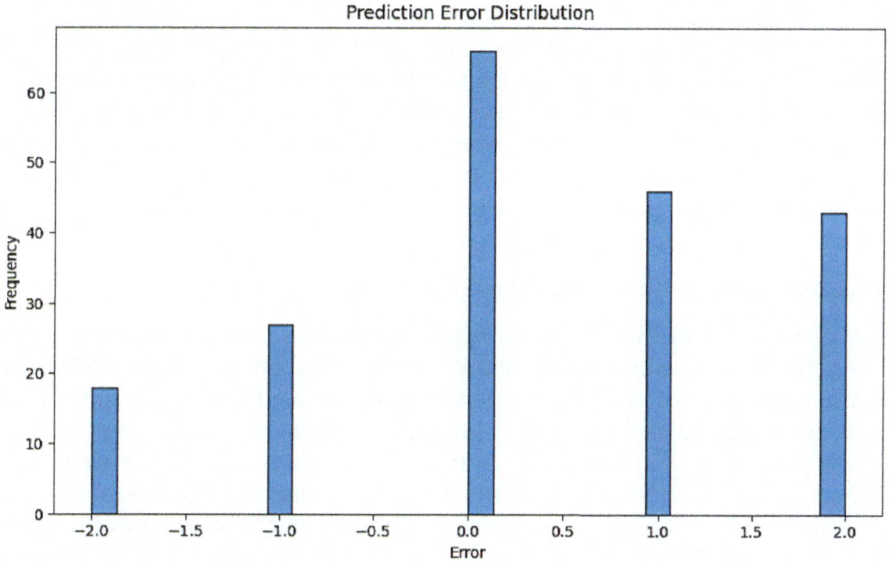

Fig. 11 Prediction error analysis

low-floor buses and level boarding platforms. TransMilenio has completely revolutionized public transit in Bogotá by carrying nearly 2 million people daily, reducing huge amounts of traffic, and air pollution. It has also motivated similar initiatives in cities like Mexico City, Johannesburg, and Istanbul, apart from serving as a model for other BRT systems.

- **London, United Kingdom—Congestion Charge**

In 2003, a London congestion charge was introduced to decrease the volume of traffic in city center areas and reduce associated air pollution. There is a daily charge for drivers who drive in the designated area; however, some types of cars, residents, and taxis are exempted from the charge. The congestion tax is payable over the internet, by telephone, or by text at specific locations. Cameras at ANPR monitor vehicles entering the pricing zone. The proceeds from the fee are put back into sustainable transportation projects such as public transit. As a result of implementing the congestion charge, there has been a 30% reduction in traffic volumes, and an improvement in air quality in central London, while traffic volumes are reduced. It has further facilitated a shift in transport modes to walking, bicycling, and public transport that make the city livable and sustainable.

- **Seoul, South Korea—Seoul Metropolitan Subway**

The Seoul Metropolitan Subway is one of the largest and most efficient systems in the world, as it passes through the capital of South Korea and the capital's vicinity. The whole area is covered by numerous lines, part of them above and part of them under the ground. A feature of multilingual announcements, computer-driven signs, and an easy-to-follow format characterizes the Seoul Subway. Another feature it is very renowned for is cleanliness and punctuality. It is interlinked with other transportation modes, including buses and commuter trains, making the transit from one line to another very convenient. Millions ride the Seoul subway every day, and it becomes an essential part of this city's transporting system, impelling urban and economic development. It has improved accessibility for locals and tourists and promoted a reduction in air pollution and traffic congestion.

- **Curitiba, Brazil—Integrated Bus System**

One of the first in the world, the Curitiba Integrated Bus System first opened in the 1970s. It makes available to the public an affordable and efficient transit system that is also very rapid because of its exclusive bus ways, platform-level boarding, and high-capacity articulated buses. The Curitiba Bus System offers smooth connections across the city, divided into many corridors, with express and feeder services. It guarantees easy accessibility with low – floor buses, raised boarding platforms, and pre-paid boarding to cut down stay times at stations. The Integrated Bus System of Curitiba has entirely reconstructed the face of transport in the city and is nowadays a global model of other BRT systems. It has created urban growth along the transport lines and concurrently reduced the traffic and pollution and increased the local's mobility.

- **Tokyo, Japan—Tokyo Metro**

One of the largest and most populated urban regions in the world, Tokyo Metro is a thorough subway system serving the Tokyo metropolitan area. Most parts of the city and its surrounding suburbs have coverage, including lines operated by several businesses. The Tokyo Metro has a high level of renown for its efficiency, reliability, and safety—the trains run at regular intervals throughout the day and are always on schedule. Modern trains, large stops with clear signs, and ample space make it easy for passengers to find their way around this large network. With a daily passenger count running into millions, Tokyo Metro also serves as the lifeline of the public transport system of the city. It has eased traffic congestion to a large extent, reduced dependence on private automobiles, and helped in environment-friendly urban development.

The following case studies illustrate the very different approaches to transportation and mobility planning and indicate what well-planned and integrated transport systems can achieve for urban mobility and economic development, as well as quality of life [58–60].

7.1 Future Trends and Challenges

Long-term foresight in transport and mobility is indispensable for developing both sustainable and efficient transport systems. If technology continues to advance, electrification and other alternative fuels can be expected to truly accelerate and see this change toward electric vehicles and charging infrastructure. If only challenges associated with technology readiness and regulatory frameworks are overcome, further, with the advent of self-driving cars, a sea change in the way transport is carried out may come about. Mobility-as-a-Service platforms are going to be game-changing to how people plan and pay for their journeys, and they will promote multimodal transportation options, ultimately making private car ownership superfluous. Large-scale urbanization and the development of megacities do not come without problems, from congestion to air pollution—underscoring the need for investments in public transit, pedestrian infrastructure, and sustainable urban development. Last-mile connectivity remains a key focus area for enhancing the efficiency and accessibility of transportation networks, with solutions like micro-mobility services and autonomous shuttles gaining traction. Data-driven decision-making processes enabled by advances in analytics and artificial intelligence will play a crucial role in optimizing transportation systems and improving user experience. Building resilience to climate change impacts and ensuring equitable access to transportation services for all members of society are ongoing priorities that require proactive policies and targeted investments. Navigating these future trends and challenges will require collaborative efforts among governments, industries, and communities to develop innovative solutions and create inclusive transportation systems that meet the needs of a rapidly evolving world [61].

Emerging technologies are poised to revolutionize the transportation sector, offering innovative solutions to longstanding challenges and opportunities for improving efficiency, safety, and sustainability. Electric vehicles are gaining prevalence through the robust increase in innovations in battery technology and charging infrastructure, which are evidenced by the governmental investments in its incentive programs to majorly reduce carbon emissions and enhance independence from fossil fuels. With autonomous vehicles, there will be great potential to improve safety on the roads and to reduce the inconvenience faced by transportation, making it accessible to any person in need. Further, MaaS platforms are picking up to deliver travelers with end-to-end planning, booking, and integrated payment functionalities for the on demand use of multimodal transportation services through the use of digital platforms. Other kinds of upcoming technologies include hyperloop systems, electric vertical takeoff, and landing—eVTOL—aircraft, and high-speed rail, coming up with utterly innovative and sustainable solutions for traveling meant to be far quicker than the conventional method of moving around from one place to another. Moreover, data analytics, artificial intelligence, and IoT technologies have been greatly enhanced with real-time monitoring, optimization, and predictive maintenance on transportation infrastructures to enhance reliability and efficiency. As these technologies continue to mature and penetrate transportation systems, the potential for change in the way people and goods move towards smarter, greener, and more accessible transportation networks will increase in the future [62].

The upheaval engineered by the trends of urbanization has changed the face of the earth as more people are gravitating to cities in search of economic opportunities, better living standards, and overall improvement in the quality of life. Today, over half of the population of the world lives in cities which are claimed to be important centers of innovation, culture, and economic activities. In more than one sense, rapid urbanization does pose a challenge and an opportunity for urban planners, policy-makers, and residents. On the one hand, they grant access to facilities for education, health care, and transport, providing the spark not only for economic growth but also for social development. Many urban areas emerge as hosts for creativity and enterprise and prompt talent and investment from around the world. However, due to this quick rate of urbanization, drawbacks include congestion, pollution, a lack of appropriate housing, and finally, social inequality. As our cities grow, the responses to these challenges lie in sustainable urban planning, the development of infrastructure, and policies that elevate inclusivity in the access to resources and opportunities in an equal manner to all. Embracing innovative solutions and harnessing the massive potential of urbanization will see the world's cities finally transform into beacons of opportunity, prosperity, and well-being for the future.

Indeed, this underlines why environmental considerations are central in providing the future directions toward transport and mobility societies that strive to mitigate climate changes, and air pollution, and protect natural resources. Much is called upon by this, given its being one of the major sectors responsible for greenhouse gases and atmospheric pollution, to shift to cleaner and sustainable modes of transport. Other strategies in the shift focus on promoting electric vehicles through public transit networks and infrastructure development in active modes such as bicycles and pedestrians, as well as shared mobility solutions like carpooling and ride-sharing. In addition, such efforts at decarbonizing transportation should aim not only at technologies in vehicles but also at improvement in fuel economy, alternative fuels, and the portioning of renewable sources into transportation systems. Doing this would also be realized by sustainable urban planning and land use policies that favor compact and mixed-use forms and reduce sprawl through transit-oriented development. Being situated on these environmental premises helps counteract the negative effects of transportation on the environment and public health, fostering the development of more resilient, livable, and equitable communities for future generations [63].

8 Conclusion

The future of transportation represents the most promising area for change, especially considering the depths of technological innovations, the needs for the environment, and the shifts in the demographic environment. This encompasses the fast pace of technological innovations surrounding EVs, AVs, and MaaS platforms. EVs promise reduced GHG emissions from transport; AVs promise increased transport safety and efficiency; MaaS platforms are promising multi-modal travel, with less emphasis on private car ownership. Making mobility less congestion-intensive and air-pollution-free requires compact, resource-efficient development in urban areas, as well as investments in public transport and infrastructure promoting active transport. However, sustainable futures for transport necessarily confront technological, regulatory, and social challenges, including effective solutions depending on structured collaboration between governments, industries, and communities. Such recommendations range from investments in renewable energy and EV recharging infrastructure to the renewal of regulatory frameworks that incentivize sustainable practices and the promotion of innovative mobility solutions through PPPs. This is also crucial in shaping public education toward more sustainable ways of travel. Deep learning integration into traffic management, as in the example of Ludhiana, showed large improvements: an accuracy result of the neural network model at 85% and validation accuracy at 82%, where high AUC in ROC curve analysis indicated that the traffic conditions were effectively discriminated. Feature importance analysis indicated vehicle count and average speed to be critical predictors. While the model did work effectively, analysis of predicted vs. actual conditions and error distribution still yielded areas for improvement. Specifically, this study contributes to a growing body of literature that has illustrated deep learning as an area that holds huge potential to better traffic management systems and suggests that similar approaches could benefit other urban areas fighting traffic woes. Future work should focus on refining models with more data and advanced algorithms to further improve accuracy and responsiveness.

References

1. Shayea, I., Ergen, M., Azmi, M.H., Çolak, S.A., Nordin, R., Daradkeh, Y.I.: Key challenges, drivers and solutions for mobility management in 5G networks: a survey. IEEE Access **8**, 172534–172552 (2020)
2. Campisi, T., Severino, A., Al-Rashid, M.A., Pau, G.: The development of the smart cities in the connected and autonomous vehicles (CAVs) era: from mobility patterns to scaling in cities. Infrastructures **6**(7), 100 (2021)
3. Gutiérrez, A., Miravet, D., Domènech, A.: COVID-19 and urban public transport services: emerging challenges and research agenda. Cities Health **5**(sup1), S177–S180 (2021)
4. Nikitas, A., Michalakopoulou, K., Njoya, E.T., Karampatzakis, D.: Artificial intelligence, transport and the smart city: definitions and dimensions of a new mobility era. Sustainability **12**(7), 2789 (2020)
5. Ahmad, F., Saad Alam, M., Saad Alsaidan, I., Shariff, S.M.: Battery swapping station for electric vehicles: opportunities and challenges. IET Smart Grid **3**(3), 280–286 (2020)
6. David, B., Yin, C., Chalon, R.: Smart coastal city mobility management in the context of high tourist activity and methodological approaches. In: The Blue Book: Smart sustainable coastal cities and blue growth strategies for marine and maritime environments, pp. 53–78. Springer International Publishing, Cham (2024)
7. Darques, R.: Traveling the country: transportation challenges and mobility management. In: The Geography of Greece: Managing Crises and Building Resilience, pp. 317–335. Springer International Publishing, Cham (2024)
8. Bellaj, M., Naja, N., Jamali, A.: Distributed mobility management support for low-latency data delivery in named data networking for UAVs. Futur. Internet **16**(2), 57 (2024)
9. Frank, L., Klopfer, A., Walther, G.: Designing corporate mobility as a service-Decision support and perspectives. Transp. Res. Part A: Policy Pract. **182**, 104011 (2024)
10. Guo, J., Chen, L., Li, L., Na, X., Vlacic, L., Wang, F.Y.: Advanced air mobility: an innovation for future diversified transportation and society. IEEE Trans. Intell. Veh. (2024)
11. Bellini, P., Bilotta, S., Collini, E., Fanfani, M., Nesi, P.: Data sources and models for integrated mobility and transport solutions. Sensors **24**(2), 441 (2024)
12. Basu, R., Ferreira, J.: Sustainable mobility in auto-dominated Metro Boston: challenges and opportunities post-COVID-19. Transp. Policy **103**, 197–210 (2021)

13. Wang, Z., Gupta, R., Han, K., Wang, H., Ganlath, A., Ammar, N., Tiwari, P.: Mobility digital twin: concept, architecture, case study, and future challenges. IEEE Internet Things J. **9**(18), 17452–17467 (2022)
14. Pan, C., Ren, H., Wang, K., Kolb, J.F., Elkashlan, M., Chen, M., Hanzo, L.: Reconfigurable intelligent surfaces for 6G systems: principles, applications, and research directions. IEEE Commun. Mag. **59**(6), 14–20 (2021)
15. Cruz, C.O., Sarmento, J.M.: "Mobility as a service platforms": a critical path towards increasing the sustainability of transportation systems. Sustainability **12**(16), 6368 (2020)
16. Lai, C., Lu, R., Zheng, D., Shen, X.: Security and privacy challenges in 5G-enabled vehicular networks. IEEE Network **34**(2), 37–45 (2020)
17. Karlsson, I.C.M., Mukhtar-Landgren, D., Smith, G., Koglin, T., Kronsell, A., Lund, E., Sochor, J.: Development and implementation of Mobility-as-a-Service-A qualitative study of barriers and enabling factors. Transp. Res. Part A: Policy Pract. **131**, 283–295 (2020)
18. Tirachini, A.: Ride-hailing, travel behaviour and sustainable mobility: an international review. Transportation **47**(4), 2011–2047 (2020)
19. Porru, S., Misso, F.E., Pani, F.E., Repetto, C.: Smart mobility and public transport: opportunities and challenges in rural and urban areas. J. Traffic Transp. Eng. **7**(1), 88–97 (2020) (English edition)
20. Butler, L., Yigitcanlar, T., Paz, A.: Smart urban mobility innovations: a comprehensive review and evaluation. Ieee Access **8**, 196034–196049 (2020)
21. Straubinger, A., Rothfeld, R., Shamiyeh, M., Büchter, K.D., Kaiser, J., Plötner, K.O.: An overview of current research and developments in urban air mobility-Setting the scene for UAM introduction. J. Air Transp. Manag. **87**, 101852 (2020)
22. Wong, Y.Z., Hensher, D.A., Mulley, C.: Mobility as a service (MaaS): charting a future context. Transp. Res. Part A: Policy Pract. **131**, 5–19 (2020)
23. Goodrich, K.H., Theodore, C.R.: Description of the NASA urban air mobility maturity level (UML) scale. In: AIAA Scitech 2021 forum, p. 1627 (2021)
24. Shaheen, S., Cohen, A.: Mobility on demand (MOD) and mobility as a service (MaaS): early understanding of shared mobility impacts and public transit partnerships. In: Demand for emerging transportation systems. Elsevier, pp. 37–59 (2020)
25. Butler, L., Yigitcanlar, T., Paz, A.: Barriers and risks of mobility-as-a-service (MaaS) adoption in cities: a systematic review of the literature. Cities **109**, 103036 (2021)
26. Paiva, S., Ahad, M.A., Tripathi, G., Feroz, N., Casalino, G.: Enabling technologies for urban smart mobility: recent trends, opportunities and challenges. Sensors **21**(6), 2143 (2021)
27. Smith, G., Hensher, D.A.: Towards a framework for mobility-as-a-service policies. Transp. Policy **89**, 54–65 (2020)
28. Tran, T., Navrátil, D., Sanders, P., Hart, J., Odarchenko, R., Barjau, C., Gomez-Barquero, D.: Enabling multicast and broadcast in the 5G core for converged fixed and mobile networks. IEEE Trans. Broadcast. **66**(2), 428–439 (2020)
29. Giordani, M., Zorzi, M.: Non-terrestrial networks in the 6G era: challenges and opportunities. IEEE Netw. **35**(2), 244–251 (2020)
30. Zijlstra, T., Durand, A., Hoogendoorn-Lanser, S., Harms, L.: Early adopters of mobility-as-a-service in the Netherlands. Transp. Policy **97**, 197–209 (2020)
31. Chen, S., Sun, S., Kang, S.: System integration of terrestrial mobile communication and satellite communication–the trends, challenges and key technologies in B5G and 6G. China Commun. **17**(12), 156–171 (2020)
32. Combs, T.S., Pardo, C.F.: Shifting streets COVID-19 mobility data: findings from a global dataset and a research agenda for transport planning and policy. Transp. Res. Interdiscip. Perspect. **9**, 100322 (2021)
33. Heidari, A., Jafari Navimipour, N., Unal, M., Zhang, G.: Machine learning applications in internet-of-drones: systematic review, recent deployments, and open issues. ACM Comput. Surv. **55**(12), 1–45 (2023)
34. Channi, H.K.: Tribological effect of 3D printing in industrial applications. In: Tribological Aspects of Additive Manufacturing. CRC Press, pp. 177–202

35. Arthurs, P., Gillam, L., Krause, P., Wang, N., Halder, K., Mouzakitis, A.: A taxonomy and survey of edge cloud computing for intelligent transportation systems and connected vehicles. IEEE Trans. Intell. Transp. Syst. **23**(7), 6206–6221 (2021)
36. Wan, S., Gu, Z., Ni, Q.: Cognitive computing and wireless communications on the edge for healthcare service robots. Comput. Commun. **149**, 99–106 (2020)
37. Nascita, A., Montieri, A., Aceto, G., Ciuonzo, D., Persico, V., Pescapé, A.: XAI meets mobile traffic classification: Understanding and improving multimodal deep learning architectures. IEEE Trans. Netw. Serv. Manage. **18**(4), 4225–4246 (2021)
38. Moscholidou, I., Pangbourne, K.: A preliminary assessment of regulatory efforts to steer smart mobility in London and Seattle. Transp. Policy **98**, 170–177 (2020)
39. Hwang, J., Li, W., Stough, L., Lee, C., Turnbull, K.: A focus group study on the potential of autonomous vehicles as a viable transportation option: perspectives from people with disabilities and public transit agencies. Transp. Res. F: Traffic Psychol. Behav. **70**, 260–274 (2020)
40. Gallo, M., Marinelli, M.: Sustainable mobility: a review of possible actions and policies. Sustainability **12**(18), 7499 (2020)
41. Kumar, P., Channi, H.K., Babbar, A., Kumar, R., Bhutto, J.K., Khan, T.Y., Wodajo, A.W.: A systematic review of nanotechnology for electric vehicles battery. Int. J. Low-Carbon Technol. **19**, 747–765 (2024)
42. Ajakwe, S.O., Kim, D.S., Lee, J.M.: Drone transportation system: systematic review of security dynamics for smart mobility. IEEE Internet Things J. **10**(16), 14462–14482 (2023)
43. Garrow, L.A., German, B.J., Leonard, C.E.: Urban air mobility: a comprehensive review and comparative analysis with autonomous and electric ground transportation for informing future research. Transp. Res. Part C: Emerg. Technol. **132**, 103377 (2021)
44. Vieira, M.A., Vieira, M., Louro, P., Vieira, P., Fernandes, R.: Adaptive traffic control using cooperative communication through visible light. SN Comput. Sci. **5**(1), 159 (2024)
45. Mutavdžija, M., Kovačic, M., Buntak, K.: Moving towards sustainable mobility: a comparative analysis of smart urban mobility in Croatian cities. Sustainability **16**(5), 2004 (2024)
46. Marino, E., Jayakrishnan, R.: Agent-based mobility simulation for potential cost and benefits of subscription services with associated cost structures. Transp. Res. Rec. **2678**(3), 207–218 (2024)
47. Yigitcanlar, T., Downie, A.T., Mathews, S., Fatima, S., MacPherson, J., Behara, K.N., Paz, A.: Digital technologies of transportation-related communication: review and the state-of-the-art. Transp. Res. Interdiscip. Perspect. **23**, 100987 (2024)
48. Kumar, P., Channi, H.K.: Optimal site selection and economic scheduling of electric vehicle charging station using HOMER: a case study of TOHANA, Haryana. In: 2024 Fourth International Conference on Advances in Electrical, Computing, Communication and Sustainable Technologies (ICAECT). IEEE, pp. 1–6 (2024)
49. Sundqvist, H., Tuominen, A.: Intermediaries and intermediation in building local transformative capacity for active and sustainable transport. Ambio **53**(1), 156–167 (2024)
50. Aposporis, P.: A review of global and regional frameworks for the integration of an unmanned aircraft system in air traffic management. Transp. Res. Interdiscip. Perspect. **24**, 101064 (2024)
51. Chen, X., Feng, S.: Age-friendly transportation policies and practices in the US and China: a comparative study. Sustainability **16**(2), 921 (2024)
52. Gijon, C., Mahmoodi, T., Toril, M., Luna-Ramírez, S., Bejarano-Luque, J.L.: SLA-driven traffic steering in B5G systems with network slicing. IEEE Trans. Veh. Technol. (2024)
53. Ahmed, F., Memon, M.A., Rajab, K., Alshahrani, H., Abdalla, M.E., Rajab, A., Shaikh, A.: Demand prediction for urban air mobility using deep learning. PeerJ Comput. Sci. **10**, e1946 (2024)
54. Nello-Deakin, S., Diaz, A.B., Roig-Costa, O., Miralles-Guasch, C., Marquet, O: Transportation research interdisciplinary perspectives
55. Kayisu, A.K., Kambale, W.V., Benarbia, T., Bokoro, P.N., Kyamakya, K.: A comprehensive literature review on artificial dataset generation for repositioning challenges in shared electric automated and connected mobility. Symmetry **16**(1), 128 (2024)

56. Biancuzzi, H., Massaro, M., Bagnoli, C.: Smart mobility in Venice: an ecosystem perspective. J. Clean. Prod. **434**, 140096 (2024)
57. Javed, M.A., Ben Hamida, E., Znaidi, W.: Security in intelligent transport systems for smart cities: from theory to practice. Sensors **16**(6), 879 (2016)
58. Ma, X., Chen, X., Dai, Z.: Special issue on transportation big data and its applications. Appl. Sci. **14**(4), 1517 (2024)
59. Zhao, S., Fan, Y., Zhao, P., Mansourian, A., Ho, H.C.: How do taxi drivers expose to fine particulate matter (PM2. 5) in a Chinese megacity: a rapid assessment incorporating with satellite-derived information and urban mobility data. Int. J. Health Geogr. **23**(1), 1–15 (2024)
60. Hazarika, A., Choudhury, N., Nasralla, M.M., Khattak, S.B.A., Rehman, I.U. Edge ML technique for smart traffic management in intelligent transportation systems. IEEE Access (2024)
61. Barrera-Jiménez, H., Pineda-Jaramillo, J.: A framework for integrating freight transport, urban land planning, and infrastructure management under economic geography principles. Urban Sci. **8**(2), 30 (2024)
62. Musolino, G.: Methods for risk reduction: modelling users' updating utilities in urban transport networks. Sustainability **16**(6), 2468 (2024)
63. Allen Ingabire, J.C., Stewart, A., Sagahutu, J.B., Urimubenshi, G., Bucyibaruta, G., Pilusa, S., Tumusiime, D.: Prevalence and levels of disability post road traffic orthopaedic injuries in Rwanda. Afr. J. Disabil. **13**, 1251 (2024)
64. Guarino, I., Aceto, G., Ciuonzo, D., Montieri, A., Persico, V., & Pescapè, A. (2024). Explainable deep-learning approaches for packet-level traffic prediction of collaboration and communication mobile apps. IEEE Open J. Commun. Soc

A Survey of Routing Protocols for Low Power and Lossy IoT Network

Anand Jyoti, Sharma Lokesh, and Shekhawat Rajveer Singh

Abstract Increasing connectivity of almost all things through the internet without any human intervention, namely, IoT, is a major trend today. To allow interactions among the things, wireless networks are being used invariably, and thus routing plays a vital role. An IPv6 based Routing Protocol for Low-Power and Lossy Networks (i.e. RPL) is given by IETF RoLL working group as a cornerstone for efficient communication among resource-constrained devices. RPL employs a hierarchical structure called Destination Oriented Directed Acyclic Graph (DODAG), making it suitable for both small scale and large scale LLN deployment. The trickle timer and objective function are the main components of RPL, which help to regulate control messages and determine the most efficient path respectively. Constraint nature of LLN leads to optimize the routing algorithm w.r.t. route discovery, route establishment, energy efficiency, network topology, scalability, reliability, latency, and error-free data transfer to the destination. This comprehensive survey focuses on a thorough examination of various RPL-based routing protocols ranges from the different routing metrics to the most recent routing solution. In this context, we also encompasses a diverse array of RPL variants, extensions, and optimizations, exploring their unique contributions, functionalities, and performance characteristics. Through an exhaustive review of existing literature and implementations, this survey aims to provide a comprehensive understanding of the evolving landscape of RPL routing protocols. The insights gleaned from this survey are poised to guide researchers, practitioners, and network architects in making informed decisions for the deployment and optimization of RPL-based solutions, ensuring robust and resilient communication in low power and lossy IoT environments.

Sharma Lokesh and Shekhawat Rajveer Singh are contributed equally to this work.

A. Jyoti (✉) · S. Rajveer Singh
Computer Science and Engineering, Manipal University Jaipur, Dahmi Kalan, Jaipur, Rajasthan, India
e-mail: jyotianand847@gmail.com

S. Lokesh
Information Technology, Manipal University Jaipur, Dahmi Kalan, Jaipur, Rajasthan, India
e-mail: lokesh.sharma@jaipur.manipal.edu

G. ALTAN and I. ABASIKELEŞ TURGUT (eds.), *Deep Learning in Ad-Hoc Wireless Networks*, Studies in Big Data 172,
https://doi.org/10.1007/978-3-031-86075-1_3

Keywords IoT · RPL · LLN · Route optimization · Routing protocols · Deep learning · Objective function · Trickle timer

1 Introduction

Primarily, the internet was used only to transfer data packets from one network point to the other using a specific IP address [1]. Later on, a large numbers of devices were enabled on the internet via wireless communication technologies such as Bluetooth, Wi-Fi, GSM/GPRS etc. This gave birth to the Internet of Things (IoT). As per [2], IoT is a fancy term where real-world objects are uniquely identified and coupled through the internet to interact autonomously. These devices are capable to sense and transmit the data from source to sink with possible analytics either at edge, fog, or cloud. Such flexibility offered by IoT has revolutionized different domains of application such as building automation, home automation, e-healthcare, smart grids, smart transportation system, industry 4.0, smart farming, etc. [3–5]. However, IoT do have some performance constraints due to limited power esp. at sensor nodes (SNs), and communication over Low power and Lossy IoT Networks (LLNs). Even, storage capacity, communication range and processing speed are rather low [6]. So, IoT is everywhere and nearly will emerge in the daily activity of human beings [7]. By 2027, it is expected that about 75 billion devices will be connected to the internet i.e., on the average nine devices for each person [8]. Resource-constrained nature of LLNs leads to higher time spent in data transmission utilizing network layer [9, 10]. An efficient routing mechanism is needed to overcome the limitations of IoT applications [11]. Once suitable protocols are discovered, we can overcome disadvantages like frequent power depletion of SNs, noise or interferences among link, heterogeneous behavior of SNs, device failures, etc. [12]. The distortion in routing pathway needs to be either reconfigured or rediscovered through an appropriate and intelligent routing strategy. That's why routing protocols should be proficient to optimize network communication addressing low power consumption. A number of solutions have been proposed and standardized by researchers in this context.

A similar concept of WSNs that existed for long, is now subsumed by IoT due to their ever increasing performance and reducing costs and size, which helped to make things being smart [13]. The tiny, tough, economical and low controlled WSNs-based SNs convey the IoT to put SNs in, on or, around the physical objects with low cost and low power [14]. As the WSNs play an important role to enable the key technologies of IoT, lots of differences exist among them. WSNs aren't directly connected to the internet, that's why data is to be sent via a router to the desired location. Hence, interoperability between WSNs and internet was very problematic task. On the other hand, IoT directly transmits the sensory data to the server due to it being enabled by of IPv4/IPv6. As per Li et al. [15], IoT performs better than WSNs w.r.t. latency and energy consumption in real time scenario. To enable IP over IEEE 802.15.4 (wireless standard), IETF (Internet Engineering Task Force) Working Group (WG) devised 6LoWPAN (IPv6 over Low-power Wireless Personal Area

Network) in 2004 [16, 17]. RFC4919, where RFC stands for request for comments, is one such document dedicated to 6LoWPAN. It came to the end in 2014, after declaration of RFC 4944, RFC 6282, RFC 6585 and RFC 6775 [18–23]. IETF also explored some well-known routing protocols of WSNs such as OSPF (Open Shortest Path First), AODV (Ad-hoc On-demand Distance Vector), OLSR (Optimized Link State Routing) etc. but did not find them suitable as none is able to sustain in LLNs environment [17]. Finally in 2012, IETF and RoLL (Routing over Low-power and Lossy networks) WG standardized IPv6 based routing protocol for low power and lossy networks e.g. RPL in RFC 6550 [24]. It makes sure to provide communication among resource constrained devices [25]. The RoLL WG addressed the routing requirements for smart urban, industrial area, home automation, and building automation applications in RFC 5548, RFC 5673, RFC 5826, and RFC 5867 respectively [26–29].

Optimized lossy network of IoT w.r.t least energy consumption is a trending research amongst researchers and applications abound in industry. Many authors perform an extensive review on routing protocols for wireless communication or LLNs. Authors in [5] gave a taxonomy of routing algorithm based on their working principle, network architecture, and application for Vehicular Adhoc Network (VANET). Comparative analysis is also done with emphasis on recent trends and direction towards reliable routing strategy. A comprehensive is carried out to highlight the traditional as well latest routing strategy with QoS in [30]. Poornima et al. focused on the algorithms which consume less energy during data communication [31]. About 120 articles have been included to conduct this review.

This survey highlights necessities of routing protocol and the design concerns for low power lossy IoT networks in Sect. 2. In Sect. 3, the routing protocols are categorized as per their characteristics and elaborated individually maintaining chronological order. A number of parameters such as mobility, traffic pattern (network topology), quality of services, energy efficiency and objective functions are considered to categorize the routing protocols. Section 4 elaborates all early routing approaches viz. 6LoWPAN, RPL and LOADng used for such networks. A large set of protocols offering improvement in one or other aspects of performance over initial proposals have appeared in literature. The same have been dealt extensively in Sect. 5. Routing algorithms over RPL in vogue are also reviewed and the section ends with a summary of gaps still left open even after latest versions, which provides insight into the future work potential. All the discussions are accompanied by comparison tables for a quick glance over the methodology, potential and weakness of each variant. The weakness of existing routing protocol specially in energy efficient routing has drawn major attention for low power lossy IoT networks and shall be foundation for our further work.

2 Necessities and Design Concerns of IoT Routing Protocols

Routing requirements may fluctuate in diverse applications of low power lossy IoT networks. As discussed above, the IETF RoLL WG presented a number of RFCs documents in this direction. Fundamentally, lossy networks are characterized by communication medium with low capacity, time variant behavior leading to high packet failure [32]. Precise characteristics of network imposes certain constraints on the design of the routing protocol. An equally important aim is to maintain the energy level for sufficient duration and upgrade the performance for any routing solution. Each protocol that is designed for LLNs only tries to satisfy a specific requirement. Maintaining all routing requirements under single solution is very contradictory task. Some challenges are identified through considering the necessities discussed in RFCs. These are listed below.

2.1 Traffic Pattern

Continuous fluctuation in network causes to be link failure, which is a major issue in designing an efficient routing algorithm. Due to this, SNs couldn't exchange the information to the desired locality. Thus, an algorithm should be able to afford distributed management of LLNs.

2.2 Scalability and Reliability

Due to limited competencies and harsh processing of SNs, routing protocols aren't able to adapt the growth of network periodically. Along with that, sometimes it couldn't deliver the data to the intended recipients profitably within a specified time period, leads to an unreliable routing strategy.

2.3 Availability

During tasks execution, network should remain operational even energy depletion of some SNs. It ensures the level of fault tolerance.

2.4 Mobility

Due to Adhoc deployment, location of devices isn't accessible easily. It is necessary to observe the current location of these devices and respond to topology changes accordingly.

2.5 Routing Loops and Convergence Time

Routing loop is a general problem, occurs in any type of network. It is a confusion regarding transmission of data from source node to the destination. A sleep cycle should be employed on routing technique, in which embedded devices only wake up when they need to communicate the data over network.

2.6 Routing Table

Proactive routing protocols are required maintaining too much information in routing table, which causes to be frequent power failure of SNs. Therefore, the routing approach shouldn't be dense with unnecessary information in the routing table.

2.7 Security and Privacy

It is one of the most important issues of LLNs, because some applications of IoT need high level of security and privacy especially in healthcare or in military. To achieve this, have to put some authentic plan through generating a key and apply it on the pathway on which data is to be transferred.

2.8 Life Span of LLN

While IoT is equipped with lots of power constraints SNs; the algorithm should restrict the consumption of energy. Many authors tried to find out a solution in this direction, still IoT is facing difficulties.

2.9 Load Balancing

The parent node or sink node lost their energy due to heavy burden in dense networking environment. Because the parent node has to transmit, receive and process the information of its all successors. It is very obligatory to balance the power consumption among SNs to get rid of packet loss.

2.10 *Heterogeneity*

As IoT is equipped with varieties of SNs and actuators, required to sense, transmit and process the heterogeneous data over the communication channel. Routing algorithm should be aware about the heterogeneity and provide interoperability among them.

3 Classification of Routing Protocols for LLNs

Routing is a procedure to choose an optimal path in the network for transferring the sensory data. A reliable routing strategy helps to conserve the energy of IoT devices. As the routing protocols of lossy IoT networks are inspired through WSNs and MANET, mainly it has three approaches such as proactive, reactive and hybrid [1]. Proactive routing protocols maintain the routing table at all points with proper entries for all destinations. Therefore, it encourages high network overhead and requires extra memory hole to process the data. While reactive routing requires less memory, and maintains the information for vigorous link only. Hybrid combines the features of both. On the whole, routing algorithm determines the lifespan of network, increase packet delivery ratio, reduce delay etc. [11]. Lots of algorithmic solutions are emerged by RoLL and IETF WG, will be discussed in Sect. 2. One of the most prominent initial routing solutions that can connect tangible devices to IPv6 is RPL [24]. Existed routing protocols create various trade-off among performances. In this section, we are presenting taxonomy of existed routing protocols. They are further classified as per their various characteristics viz. P2P, mobility, energy efficiency, trickle timer, objective functions etc., shown in Fig. 2. P2P is the primary need of any IoT applications and the RPL follow same [13]. The routing protocols categorization features are discussed as below [7].

3.1 *Peer-to-Peer Routing Protocols*

Most IoT scenario needs to support peer-to-peer traffic pattern for data communication. Default RPL follows this, which is devised by IETF. Subsequently, RoLL WG proposed P2P-RPL in 2013. It is a reactive approach. P2P-RPL, GeoRank and ER-RPL etc. are come under this category.

3.2 Mobility Based Routing Protocols

In order to perform some significant tasks with multicast transmission, mobile SNs are needed. Industrial-IoT and e-healthcare must equip with mobile SNs for continuous monitoring and reliable data transfer. Therefore, mobile nature of SNs should be sustained by routing protocols.

3.3 Diverse Traffic Pattern and Mode of Operations Based Routing Protocols

During data processing, IoT applications require different traffic pattern. When two SNs communicate to each other need P2P communication, but one of them needs to communicate with the sink using MP2P. Mode of Operation (MOP) checks whether the SNs are in storing or non-storing mode. Hence, routing algorithms should support different traffic pattern to enhance network performances.

3.4 Energy Efficient Routing Protocols

Emerging IoT applications need quick response and efficient data transfer. It is needed to transfer the data packets with minimum power consumption, because SNs are equipped with less energy. Recharge or replace the SNs frequently are also tough task.

3.5 Routing Protocols Based on Objective Function

An objective function (OF) states that the result of the process utilized by SN to choose and upgrade routes within RPL network. For this, it computes the rank of the device within the DODAG version. Fundamentally, the OF concludes procedure to select the best preferred parent node.

3.6 Security Aware Deep Learning Based Routing Protocols

It's possible to control devices remotely due to the sensor and communication network via an internet connection [33]. During packet transmission among various IoT devices, has the chances of vulnerability. Hence, security is a major concern of LLN.

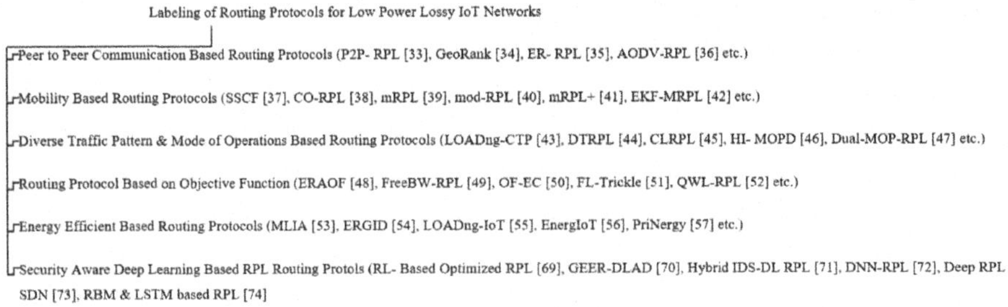

Fig. 1 Taxonomy of studied routing protocol in IoT

Deep Learning has the capability to evaluate the concerned Cybersecurity domains based on collected data from them [34] and draw some meaning insights (Fig. 1).

4 About RPL

As discussed above, 6LoWPAN, RPL and LOADng are initial routing solutions for low power and lossy IoT networks facilitated by IETF RoLL WG. They are aiming to enable IPv6 among IEEE 802.15.4 i.e., low power devices. RPL refers to an IPv6 based Routing Protocol for Low power and lossy networks (RPL), came in 2012 [24]. It is a tree based proactive routing, maintains Destination Oriented Directed Acyclic Graphs (DODAGs) for creating the routing pathway. RPL uses four identifiers RPLInstance_ID, DODAG_ID, DODAG-Version_Number and rank to maintain its network topology. Set of one or more DODAGs are identified with RPLInstance_ID. Each DODAG consists of a root or sink (also known as LLN Border Router) that is identified by unique DODAG_ID identifier. DODAG-Version_Number is incremented whenever reconstruction of DODAG is required. Rank helps to notify the location of SN w.r.t. the root of DODAG and also avoids the loops. RPL defines the objective function (OF) to optimize the DODAG without containing any loop. For this, an OF calculates the best pathway using routing metrics hop count, ETX (Expected Transmission Count) and energy of SNs. The key motive of RPL is to reduce routing signals, energy consumption and complexity. Parameter control technology restricts the frequency and message count in network control management system [35].

Numerous control messages viz. DIO (DODAG Information Object), DAO (Destination Advertisement Number) and DIS (DODAG Information Solicitation) have been used in RPL for network continuation. Initially, the sink node starts to send the DIO message to its neighboring nodes. The DIO contains the relevant information such as operation mode, node rank, and unique identification and OF. After receiving this DIO message, the receiving nodes decide whether to join the DODAG or not. If the nodes desire to join, send DAO message to its parent node or sink node through upward routing pathway. During this, leaf node updates its routing table and chooses a preferable parent node for further data transmission to the DODAG root. The parent node is chosen with minimum rank, calculated as per the OF. If SN is considered as a router, should update and resend the DIO message to its neighbors otherwise considered as leaf node. RPL allows a new SN to join the DODAG anytime through sending a request using DIS (DODAG Information Solicitation) for DIO. When SN receives DIS, join the DODAG and choose parent using OF. Furthermore, Trickle algorithm helps to improve the network stability by avoiding redundant information w.r.t DIO.

To maintain stability of RPL network, the DIO message is circulated among SNs on the basis of trickle timer algorithm [36]. Imin, Imax and Idoubling variables configure this trickle algorithm [11]. Imin and Imax are the minimum and maximum interval size of trickle size of the trickle timer respectively. Idoubling is a decision parameter for Imax in RPL. As diverse applications of IoT don't require same thing, it's very crucial to choose appropriate values for these three variables. Higher value of Imax improve energy efficiency but less responsive and it's vice versa applies for the value of Imin.

RPL permits both upward and downward route from the SNs to its parent node and its children nodes respectively, shown in Fig. 2. Upward route helps to manage

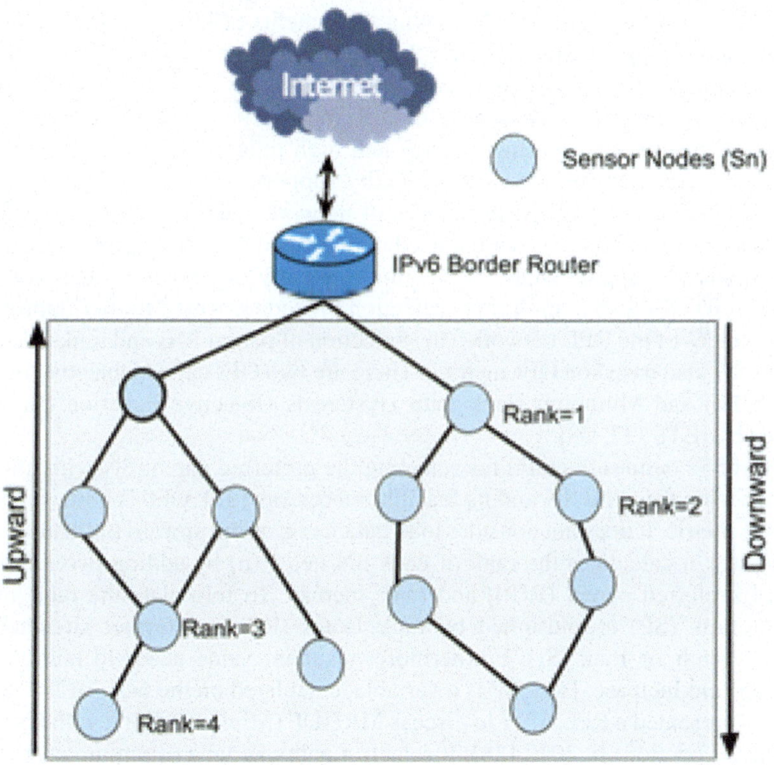

Fig. 2 DODAG formation in RPL network

the DIO messages during network creation while downward route controls all DAO messages. This DAO control messages are required to acknowledge with DAO-ACK control message by receiving SNs. It is an optional feature for an application of IoT that need P2MP and P2P traffic pattern. Downward routes work in both storing and non-storing mode. In both scenarios, flow of P2P data packets in up and down direction towards DODAG root and desired location respectively.

Three types of traffic patterns i.e., multipoint-to-point (MP2P), point-to-multipoint (P2MP) and point-to-point (P2P) are carried in RPL. In P2P, the root SN sends the data packets to the destination using upward and downward flow. Most of LLNs required this pattern. P2MP uses destination advertisement mechanism, in which root sends the data packets in downward direction to the other SNs. It's also called multicast transmission. While in MP2P, SNs send the data packets toward DODAG root in upward direction.

RPL facilitates different modes of operations (MOPs) for diverse applications of IoT. MOPs are selected as per the required traffic pattern and computational capability of SNs in the network. Four MOPs i.e. MOP0, MOP1, MOP2 and MOP3 are there. No downward routes are found in MP0, only MP2P traffic pattern is allowed here. MP1 works in non-storing mode, which uses P2P and P2MP traffic pattern for communication in downward routes. DODAG root maintains the information of all downward routes of RPL for further transmission of data packets. In MOP2, each SN maintains its own routing table by taking the benefits of DAO message. This DAO is circulated during downward traffic. MOP3 supports storing mode with multicast transmission. It allows a non-root SN to send data packets to a group of SNs in RPL.

As discussed above, performance of RPL may vary in various scenarios of IoT. Suppose E-healthcare need low latency and high reliability. While other domains need low energy consumption with mobility supports. Here, OF plays a vital role to upgrade or degrade the performances of network. Metrics of OFs are varied in each scenario and chosen as per the need of dedicated network performances. Many reasons behinds this. (i) DODAG creation is greatly dependent on the OFs of the network. (ii) OFs also help in route selection procedure which further improves the performances of the RPL network. (iii) Selection of parent SNs and rank calculation of each SN also based on OFs metrics. There are two OFs called Objective Function Zero (OF0) and Minimum Rank with Hysteresis Objective Function (MRHOF), devised by IETF [37, 38].

RFC 6552 defines the OF0 for selecting the preferred parent SN with minimum distance from the child SN and its feasible successors [37]. OF0 is quite simple and uses one metric. It does not consider load balancing, as the upward traffic uses parent SN. Firstly, it calculates the rank of each SN i.e., R(n) by adding two metrics the rank of preferred parent (R(P)) and rank_increse. To formulate the rank_increse, step_of_rank (SP) is multiplied by rank_factor (RF) and further stretched with a term stretch_of_rank (Sr). Furthermore, resultant value needs to multiply with MinHopRankIncrease. Here SP is a variable, calculated on the basis of link quality.

IETF presented a RFC 6719 to discuss MRHOF (Minimum Rank with Hysteresis Objective Function) in 2012 [38]. It selects a pathway with minimum cost without promoting excessive changes of the preferred parent. To accomplish this, RPL first select routing pathway then perform hysteresis process. A candidate parent node is chosen as preferred parent when the cost of its pathway is smaller than the existing parent during this process. ETX metric is taken to compute the pathway cost. Many numbers of authors studied about RPL and proposed detailed survey in [39, 40].

5 Improvements over RPL Routing Protocols

At current, most of the proposed routing algorithms are the improvement over RPL. Numbers of surveyed have been done towards this direction by different authors in their own way [10]. Taxonomy of studied routing protocols is revealed in Fig. 2. An extensive review of each is provided in the chronological order, which helps to understand very well. In coming subsections, protocols have been analyzed as per their characteristics. At the end of each subsection, protocols are summarized in a table based on their methodology, features and drawbacks. The methodology column discusses the algorithm to achieve the goal of proposed work. The features and drawbacks columns tell the contribution and fault respectively of each protocol.

5.1 Peer to Peer Communication Based Routing Protocols

The fundamental approach of RPL couldn't meet with the requirements of building automation and remote-control areas. These applications of IoT must need to support P2P communication [64]. Then IETF and RoLL WG have started to enhance the RPL to support P2P communication. Finally, in 2013 they proposed a Point-to-Point router discovery mechanism over RPL i.e., P2P-RPL) [41]. It is reactive in nature, discovers the best pathway among source and destination in on-demand fashion. When the source SN is required to send the data, creates a temporary DODAG and became the parent SN. For this, the source SN starts route discovery process using trickle algorithm. At first, it circulates the control message P2P-RDO (P2P Route Discovery Option) throughout the network. The receiving SNs check the message destination and determine whether to join the DODAG or not. If it doesn't join, forwards the P2P-RDO across the network until it reaches its destination SN. After receiving, the destination SN must generate the P2P Discovery Reply Object (P2P-DRO) message to reply the route request. Then receiving SNs ensure availability about the source node enfolded in P2P-DRO. If it isn't, the nodes are meant to be intermediate nodes and further send this message to the preferred parent node. The P2P-DRO is required to forward back, so that communication between source and destination starts. Here i.e., in P2P, lifespan of this provisional DODAG depends on the lifespan of requested parent SN. In [42], authors explored a new solution to minimize the probability of connectivity loss in RPL. It finds an alternative parent node using DIS message and link estimation function.

Barriquello et al. has projected IPv6 enable geographic routing approach for LLNs, called GeoRank [43]. This routing approach is the improvement over RPL, with using the GOAFR (Greedy Other Adaptive Face Routing) algorithm. The GOAFR is a location aware routing algorithm, able to find an efficient pathway in high density network area like smart city. At first, GeoRank find out the distance between the sources SNs to the destination for all current roots of DODAGs. Then, one root is selected as anchor SN with the lowest absolute angle difference between the source and destination. Furthermore, it forwards the data using greedy forwarding approach where the SN sends data to the intermediate SN in single hop. Intermediate SN further transmits the data to the destination SN. If it fails, GeoRank mode sends the data to the preferred root using anchor SN. This process should be going on until the data reaches to the SN closer to the destination than the anchor SN. Else data arrives to the anchor for further transmission using GOAFR. As the GeoRank needs geometrical calculation, SNs are static and must be equipped with GPS. It has some mobile SNs but must be one hop away from the static SNs. Finally, GeoRank also able reduce the number of control messages, because leaf SNs doesn't need to transmit DIO to acknowledge the source SN.

Although RPL and P2P-RPL are flooded with number of control messages, this leads to network overhead and power depletion problem. In [44], Zhao et al. has tried to give an energy efficient routing protocol for LLNs without losing their reliability. This is named as Energy-efficient Region-based RPL (ER-RPL) routing protocol. Combining the proactive feature of RPL and reactive feature of P2P-RPL, it divides the network's region into numbers of provinces and starts to discover the route. Like GeoRank, ER-RPL also uses the SNs that are equipped with the GPS. These SNs are called reference nodes i.e., RNs, help in network partitioning and distance calculation (from SNs to the RNs). To identify each network province, a Region Code (RC) is assigned. At the initial phase of route discovery, default parent of RPL circulates a Message Request Object (MRO) message to its target SNs. It checks the routing pathway is valid or not based on the route cost. If it is, inform to the parent else starts to discover the route based on the network province. During this, the target SNs create a provisional DODAG using MRO and start to send the data packets. At this point of time, RNs are sitting idle. Hence, ER-RPL is capable to send the data using the target SN and achieve reliability. Energy consumption is also diminished because numbers of control messages are reduced.

It can be seen that P2P-RPL supports only asymmetric route, which isn't suitable for large network area. Anamalamudi et al. proposed Ad-hoc On-demand Distance Vector RPL based routing (AODV-RPL) in 2018 [45]. It is a P2P based reactive routing approach, which endorses both symmetric and asymmetric routing path as well. AODV-RPL creates route only when it's really needed i.e., on demand basis. The source SN starts to discover the pathway, either if existing one isn't able to satisfy the requirement of concerning application or the previous pathway doesn't exist. For this, an RREQ-DIO message is sent to the target SN for creating RREQ instance. Here, RREQ-DIO comprises the info (S field) about the pathway is symmetric or asymmetric. The receiving SNs join the RREQ instance after checking the condition for bidirectional pathway i.e., $S = 1$. If it joins, the pathway is considered

Table 1 Comparison among P2P routing protocols of LLNs

Routing protocol	Author's name	Methodology	Features	Limitations
P2P RPL, 2013 [41]	Goyal et al.	Uses reactive approach for route discovery and establish P2P route for data communication	Stability period is increased, because the parent SN doesn't involve during P2P data communication	Increase network overhead and power consumptions due to circulating numbers of control messages
Geo Rank, 2015 [43]	Barriquello et al.	Integrates RPL and GOFR for route maintenance and data packet delivery resp.	Few no. of control messages are used and more scalable	SNs are static and must be equipped with GPS
ER-RPL, 2016 [44]	Zhao et al.	Combines the features of RPL and P2P-RPL and divides the network area into number of provinces	Reduce routing overhead and achieve reliability as well as energy efficiency	complex approach because added one more control message
AODV-RPL, 2018 [45]	Anamalamudi et al.	Symmetric route is used for creating DODAGs	follows P2P pattern and Shrink the size of control messages	Cause to be frequent link failure

as symmetric for remaining all that will use this link. After receiving the message RREQ-DIO, the target SN decides to send RREP-DIO to the source SN via created symmetric pathway. If 'S' bit is set to 0, AODV-RPL considers it asymmetric. In this case, the target SN creates DODAG of RREQ to multicast RREP-DIO its all neighboring SNs. Hence, AODV-RPL creates a pair of DODAGs in single route discovery process rather than finding a symmetric path between SNs (Table 1).

5.2 *Mobility Based Routing Protocols*

In 2016, authors are investigated a Sink-to-Sink Coordination Framework (SSCF) to enhance RPL [46]. Coordination is achieved by circulating the periodic route maintenance messages (like DIO and DAO) among the sink SNs. Although SSCF assumes multiple static or mobile sink within same network, that's why more than DODAGs are needed in DAG. Each sub-DODAG is identified through the unique id of sink node itself, and broadcast a DIO message to its all neighboring SNs. The receiving SNs update their rank and decide to join the DODAG accordingly, known as border or edge SNs. Then DAO message is used by receiving SNs to notify that they have joined the network. SSCF also appends a network coordination interval (NCI) to define an interval in which every edge SN receives an updates regarding Network Optimization Metrics (NOMs) status. The value of NCI should be less than DIO interval and varies in different domains of IoT. The parent table takes responsibility to manage the subDODAG-ID and NOMs values, encapsulated in DAO. Simulations show that, SSCF performs better w.r.t. throughput, end-to-end latency and power consumption than RPL.

In order to maintain the QoS of LLNs, Gaddour et al. proposed Corona mechanism aware RPL (Co-RPL) routing protocol [47]. It divides the whole network area into number of circular areas i.e., coronas with fix radius. Each SN is responsible to maintain a routing table for tracking its neighboring SNs with four control messages such as SN-ID, DAG-ID, CORONA-ID and link quality. Here, a DAG-ID is assigned to each corona. A SN chooses its parent node and defines its CORONA-ID, after receiving the DIO message. To evaluate the CORONA-ID, one is added to the lowest CORONA-ID among its neighboring SNs. CORONA_ID also detects the location of mobile SNs in respective DAG. Co-RPL is able to find an alternative parent in case of connection loss as the next hop. For this, SNs circulates DIS message by adding a new bit F as the first bit. It notifies the leaves SNs to stop the sending of data packets. This interaction happens until the switch hub finds another favored parent. Performance of network is analyzed on the basis of packet loss ratio, power consumption, average hop count and end-to-end delay.

In [48], authors incorporate a hand-off mechanism (smart-hop) into the existing RPL to bear the mobile SNs. It is very simple, effective and backward compatible with the de facto protocols of LLNs i.e., RPL. Due to smart-hop, mobile SNs easily switch from one place to another. Four timers i.e. connectivity timer (Tc), mobility detection timer (TMD), hand-off timer (THO) and reply timer (TR) are used during the process of data transmission and data discovery. During data transmission, the mobile SN sends a DIS to the access point (AP). In turn, AP calculates average received signal strength (ARSS) and sends a DIO message. If the value of ARSS is below than the defined threshold, discovery phase is started with a new parent. Here the control messages of RPL viz. DIS and DIO are used as beacons. mRPL avoids the collision as well as loop creation during data collection by using the mobile SNs.

To support the mobility especially in healthcare domain, authors introduced modify RPL (mod-RPL) [49]. It is a hybrid approach and uses static as well as mobile SNs. As mod-RPL is an extension of RPL, all leaves SNs of DAG are worked as mobile SNs. Here mobile SNs aren't permissible to send DIO control messages. A parent SN is selected among mobile SNs with the minimum path cost. The operations performed by static nodes are same as traditional RPL. Hence, mod-RPL avoids the creation of sub-DODAG and connection loss in the network.

To achieve interoperability among static SNs and mobile SNs in IoT environment is a major issue. Hence, authors tried to avoid this issue in [50]. This approach is named as mRPL+ that integrates the concept of mobility with RPL and 6LoWPAN stack in backward manner. Basically mRPL+ is a proactive routing algorithm. It uses two hand-off mechanisms (i.e. hard and soft) to boost up the mobility in IoT. Hard hand-off mechanism disconnects the link between a mobile SN and AP before the formation of new link. While, soft hand-off mechanism establish a new link before smash the current link. Most of the mobility aware routing protocols degrade the performance of network due to the mobile SNs. But mRPL+ solves this issue by detecting the mobile SNs on the basis of link quality as well as parent SN unreachably test. It has a dynamic timer to check the mobility, which reduces the network overhead. Simulation results show that mRPL+ achieve 100% throughput while RPL achieve only 80%.

Aiming to support mobility in RPL, Bouaziz et al. proposed Extended Kalman Filter RPL (EKF-MRPL) routing [51]. It is a proactive routing. EKF-MRPL is consisting of three stages i.e. movement detection phase, reaction phase and notification phase. Data collection procedure remains same as RPL except mobile SNs. In first phase, each mobile SN has its own preferred parent (PP) and receives DIS message from the PP. The PP is responsible to identify the movement of mobile SN. For this, PP calculates the RSSI value and compares with given threshold. If it is less, the PP assumes the mobile SN has leaved its coverage area. Second phase is all about to gather the information and select a new PP. Here, the location of all current mobile SN is to be estimated and predicted through Extended Kalman Filter. At the last phase, all new routes are updated after choosing a new PP by circulating the message 'DIO'. Henceforth, EKF-MRPL is able to reduce the signaling cost and power consumption of LLNs (Table 2).

5.3 Traffic Diversity and MOP Based Routing Protocols

The modified version of LOADng, which is able to adapt the different structure of network and data link layer features referred as LOADng-CTP [52]. This is devised by Yi et al. It appends some additional perception in LOADng: (i) 'smart route request (SmartRREQ)' and 'expanding ring search' to improve the efficiency of route discovery process, (ii) a 'collection tree protocol (LOADng-CTP)' for diminishing the routing overhead, (iii) the 'depth first forwarding (DFF)' to recover from the data packet breakdown. In case of SmartRREQ, a router starts to discover a route by broadcasting a message with smart-rreq flag i.e. RREQ_SMART. This extended version of LOADng facilitates security because only truthful router forward the data and send RREP message to the instigator of RREQ in case of malicious surroundings. RREQ_Trigger, RREQ_Build and HELLO messages were introduced to discover a bidirectional link, establish a route to the sink and verify the bidirectional link respectively by LOADng-CTP. DFF suggests a new route in case of link breakdown to send the data packets using heuristic method.

As it is mentioned above, RPL is to be considered as de-facto routing protocols for LLNs. It doesn't support diverse traffic pattern. To avoid this problem, authors have been proposed an efficient approach in [53]. This is named as DT-RPL i.e. Diverse-bidirectional Traffic-delivery through RPL. Basically DT-RPL has three stepladders i.e. link estimation (ETX) of parent SN, delivery of ETX to the child SNs and the updation of child link quality. In this, the parent SN and its all child SNs are considered

Table 2 Mobility aware routing protocols of LLNs

Routing protocol	Author's name	Methodology	Features	Limitations
SSCF, 2016 [46]	Khan et al.	Introduces coordination framework to maintain the network topology	performs better w.r.t. throughput, end-to-end latency and power consumption	Needs to maintain two extra values in their parent table, leads to overhead
Co-RPL, 2014 [47]	Gaddour et al.	Uses corona mechanism to divide network area and also added extra bit in control message of RPL	Improve network performances w.r.t. lifespan, network latency, power consumption and link connectivity etc.	Routing table is still flooded with lots of info. and assume only static sink
mRPL, 2014 [48]	Fotouhi et al.	Incorporate a smart-hop mechanism with existing RPL	Avoids collision and loop creation. Hence, increase throughput and efficiency of network	Flooded with the number of exchanged control messages
MOD-RPL, 2015 [49]	Gara et al.	All leaf SNs of DODAG are considered as mobile SNs	Easily incorporated with default RPL and avoid the creation of sub-DODAG	Prohibits the use of mobile SNs as a router
mRPL+, 2017 [50]	Fotouhi et al.	Two hand-off mechanisms (i.e. hard and soft) are used to enhance mRPL.	Provide interoperability among static and mobile SNs	Data collection rate is low
EKF-MRPL, 2017 [51]	Bouaziz et al.	Uses extended Kalman filter to find SNs' location	Able to reduce the signaling cost and power consumption	Doesn't scalable i.e. needs to investigate w.r.t. end to end delay

as sender and receiver respectively. The data packets carry the information of link quality inside the IEEE.802.15.4 header. Only parent SN is allow for computing the link quality in downward data transmission and further transmitting to the child SN. There are some reserve spaces in the header of IEEE 802.15.4 to acknowledge the child SN about link quality. DT-RPL achieves the updation of fast link quality in both upward as well as downward by avoiding the additional overhead. Simulation shows that, DT-RPL has better performance w.r.t. throughput (packet delivery ratio), control overhead and duty cycle than RPL.

Due to the heavy load on the network, RPL didn't fit in most of LLNs atmosphere. Heavy load leads to high rate of packet loss and power depletion. Taghizadeh et al. devised a solution to avoid these problems, called Context-Aware and Load-balancing

RPL (CLRPL) [54]. This routing algorithm works in three phases, all are directly linked and cooperative to each other. At the initial phase, an objective function i.e. CAOF (Context Aware Objective Function) calculates the rank of each SN in the network. Rank calculation considers the parameters viz. ETX, residual energy and rank of the parent SN. CAOF gradually shifts the high rank value to the real rank value, and eliminates the Thundering Herd Phenomenon. The next phase introduces an efficient routing metric, called Context-Aware Routing Metric (CARF). It focuses on residual power and queue utilization of parent chain on the way to root in recursive manner. This is considered as a decision parameter for choosing the parent SN. Lastly, a new mechanism is given by authors to select the best parent SN in the dense network area.

In order to raise the reliability in downward communication of RPL, Lee et al. devised HI-MOPD (Hop Interval based Mode of Operation Decision) algorithm [55]. For this, it characterizes two MOPs for intermediary SNs i.e. storing mode and non-storing mode. The SNs which are in storing mode require space in memory to update the information of source routing along with local procedure. While the non-storing SNs involve in packet relay operation rather than consuming the memory space. The sink SN estimates optimal hop interval for storing SNs (OHIS) and sends a 'DIO' message to its all intermediary SNs. This estimation decides the number of storing and non-storing SNs in the network. After receiving the DIO, the intermediary SNs compares the OHIS from itself to the closest storing SNs. If it is nearer to the estimated hop interval, considered as non-storing SNs and save the memory space. Else the intermediary SNs are still considered as storing SNs. Hence, HI-MOPD has greater performances in terms of packet delivery ratio and path recovery than traditional RPL. Importantly, this routing algorithm is designed for only Underwater Wireless Sensor Networks (UWSNs).

It has been seen that, traditional RPL goes through a connectivity issue when two MOPs are combined on a single network. Hence to support the multiple traffic pattern in downward routing of RPL, authors have been suggested a lightweight solution in [56]. This solution is named as DualMOP-RPL, which uses both storing and non-storing SNs. Authors modified the RPL by introducing five major developments. The first development avoids the network partitioning and considers all leaf SNs as routers. In second, this algorithm allows the non-storing SNs to transmit the DAOs in hop-by-hop fashion. On the other side, storing SNs store and process the DAO messaged received from storing as well as non-storing SNs. The third and fourth developments solve the issue of source router header problem through modification of DAO messages. Fifth development is considered as optional, because it informs the DODAG root uses storing SNs or non-storing SNs. There are various overlapping networks under same region for obliging the single dense network having different MOPs. Therefore, DualMOP-RPL improves link quality, routing performance and network migrations of the LLNs. It also achieves higher throughput with bidirectional communication (Table 3).

Table 3 Traffic and MOP routing protocols of LLNs

Routing protocol	Author's name	Methodology	Features	Limitations
LOADng-CTP, 2014 [52]	Clause et al.	Build a collection tree protocol with smart RREQ supporting diverse MOPs	Reduce network overhead and compose bidirectional route	Needs additional data structure, so occupy more space in memory
DT-RPL, 2017 [53]	Kim et al.	Consider variant traffic pattern and compute ETX for parent and child SNs	Link quality has been improved in upward and downward traffic	Limited for specific domain of IoT or LLNs
CLRPL, 2018 [54]	Taghizadeh et al.	Introduce new routing metric using CAOF, CARF and parent selection mechanism	Improve throughput along with frequent power depletion issue	End-to-end delay is increased now and also needs extra memory space
HI-MOPD, 2017 [55]	Lee et al.	Proposes a MOP decision procedure to classify the storing and non-storing SNs	Achieve high level of QoS w.r.t. throughput and path recovery	Suffers from memory constraints due to its various storing SNs
DualMOP-RPL, 2015 [56]	Ko et al.	Modifies the various control messages of RPL along with using storing and non-storing mode of SNs	Fill the gap of interoperability among different modes of operation	Control messages need to process more information

5.4 Routing Protocol Based on Objective Function

In [57], Sousa et al. invented an objective function to improve the reliability and efficiency of RPL. This novel method is known as Energy Efficient and Path Reliability Aware Objective Function (ERAOF). Two network parameters i.e. Energy Consumption (EC) and Expected Transmission Count (ETX) are used to optimize the route selection procedure. EC evades the path with low energy level, while ETX aims to establish the pathway with maximum energy. Chances of link failure are very low. Firstly DIO control message is circulated around the network after calculating the link quality (T(ni)) of each SN. T(ni) is the sum of FEC(ni) and FETX(ni, nj). FEC(ni) denotes energy consumption of particular SN 'i'. Whereas FETX(ni, nj) return ETX value of the link between a SN 'i' and the DIO message sender 'j'. Now, SN 'i' forwards DIO to its all successor with the sum of calculated value plus T(nj). The SN 'j' is the previously received within DIO. Next, ERAOF selects the best pathway based on the computed value Q(r). It is the sum of T(ni) values for all SNs. In this way, it decreases the use of connections and improves the performance of LLNs.

To enhance the efficiency of data communication in multimedia environments of LLNs, a new OF has been introduced [58]. This is the enhanced version of RPL, called Free Bandwidth RPL (FreeBW-RPL). It's able to sense the data of various multimedia devices. A parent SN is chosen through taking the consideration of free space of maximum bandwidth. FreeBW-RPL calculates the bandwidth at the network layer rather than MAC layer. It is obtained by subtracting the Sendrate (used bandwidth in routing procedure) from Linkrate (total number of packets). FreeBW-RPL also avoids the highly congested pathway by switching the path into a new parent SN with less congestion. Therefore, it outperforms w.r.t. throughput, end-to-end delay and power utilization.

Aiming to fulfill the necessities of LLNs, authors investigated OF-EC (objective function based on the combined metrics ETX and EC along with fuzzy logic method) [59]. They used set of metrics and constraints viz. ETX, HC (Hop Count) and EC based on fuzzy logic. It allows converting the set of metrics into single. ETX finds the number of retransmission needed to receive the data packets at destination. The product of forward (df) and reverse (dr) delivery ratio is the way to calculate ETX. EC is total energy spent by SNs during data transmission. Here, the membership function of ETX and EC detects the quality of pathway. The minimum values of ETX and EC assist an excellent pathway. Performances are measured in two different topologies i.e. random and grid. Result reveals the superiority of OF-EC w.r.t. packet delivery ratio and network overhead over RPL.

To upgrade the performance of RPL, author enhanced the OF along with trickle algorithm called FL-Trickle [60]. Generally, trickle algorithm controls and regulates the transmission of data packets in the network. FL-Trickle modifies the value of 'T' (time) and 'I' (interval). T is the time to transmit the control traffic. It is needed to decrease the value of 'T' by I/2 to make network convergence fast. High value of Imin decreases the network overhead. Then the RPL-EC (RPL based combined ETX and energy consumption) and RPL-FL (RPL based on flexible trickle method) are compared. As a result, RPL-FL increases throughput and decreases energy consumption. However, RPL-EC has better convergence time and network lifespan than RPL.

The OF0 and MRHOF of RPL don't well suit in heterogeneous traffic pattern of IoT. Hence, Musaddiq et al. delivered a new routing approach called Queue and Work Load RPL (QWL-RPL) [61]. It tries to solve the load unbalancing problem in various network patterns. Periodic workload information (total number of transmission in every 10 s) of SNs is used for load balancing. The new objective function of QWL-RPL is approached by two parameters i.e. congestion status and link workload of SNs. It evades using the ETX metric in new objective function due to large amount of network overhead and delay. Henceforth, this routing protocol improves the overall performance of network along with providing a reliable pathway.

To maintain a reliable and optimal pathway, Hatem et al. has modified the MRHOF in [62]. This objective function considers a combination of metrics viz. power, hop-count, ETX, throughput and latency. ETX calculates the path cost from the SNs to the parent node. This modified approach integrates the value of energy and ETX with MRHOF.

In [63], Hassani et al. stated Forwarding Traffic Consciousness Objective Function (FTC-OF) RPL routing for heavy traffic lossy network. Three routing metrics are used here, naming as hop count, RSSI (Radio Signal Strength Indicator) and Forwarded Traffic Metric (FTM). FTC-OF allows to select the shortest pathway with link quality metric RSSI. Hop Count (HC) tells about the total numbers of SNs in the particular pathway (between candidate and sink). If HC(SN) equals to 1, considered as sink node. Next, the value of RSSI is obtained by the addition of RSSI-register and RSSI-offset at the MAC layer. Then each SN finds out its RSSI value to acknowledge DIO message. FTM counts the numbers of data packets created locally or to be forwarded, done on the application layer. After processing these three metrics rank of each SN (R(SN)) is calculated. It is the sum of rank of candidate SN (R(c)) and rank_increase. Furthermore, the rank_increase is the combination of IHC(SN), RSSI (SN) and FTM(SN) with additional multiplicative factor α. Here, α adjust the value of FTM. Simulation shows that, FTC-OF increases the network reliability, diminishes the power consumption and balances the workload among SNs in the network (Table 4).

5.5 Energy Efficient Based Routing Protocols

Li et al. devised a hierarchical clustering method based on duty cycle concept called EnergIoT [64]. Different duty cycle ratio helps to increase the lifespan of power constraint IoT devices. It allows to switch the devices between two different modes i.e. sleeping and active. In active mode, SNs consume power for sensing, receiving and transmitting the data. Whereas sensing function is turn off during sleeping mode. EnergIoT estimates power consumption of IoT devices during network formation as well as data processing. Therefore, it's able to improve the network lifespan without sacrificing the performances of network (i.e. end-to-end delay).

Valls et al. has been jointly improved the life span of network and performance of data analytics jobs in [65]. It is naming as maximum lifetime IoT analytic (MLIA), which resolves the issues without knowing the ability of time-varying network link. MLIA provides an online platform to execute convex problem and routing policies in support for data analytics. Data can be processed (i) locally, (ii) at the gateway or iii) anywhere in the IoT.

In [66], Qiu et al. tried to improve the reliability as well as efficiency for quick response of events in IoT. This methodology is termed as Emergency Response IoT based on Global Information Decision (ERGID). It consists of two methods i.e. Delay Iterative Method (DIM) and load balancing forwarding method called REPC (Residual Energy Probability Choice). DIM helps to alleviate the quandary

Table 4 Routing protocols based on OFs in LLNs

Routing protocol	Author's name	Methodology	Features	Limitations
ERAOF, 2012 [57]	Sousa et al.	Uses two parameters i.e. EC and ETX to optimize the route selection procedure	Achieve high throughput along with less connections	Doesn't support highly dense network area of LLNs
FreeBW-RPL, 2020 [58]	Bouzebiba et al.	Calculate bandwidth at network layer rather than MAC layer	Improves end-to-end delay, throughput and network lifespan	Chooses a longer pathway for high bandwidth
OF-EC, 2017 [59]	Lamaazi et al.	combine set of metrics viz. ETX, hop count and EC into single using fuzzy logic approach	Improve packet delivery ratio and reduce network overhead	consumes more space in memory
FL-Trickle, 2019 [60]	Lamaazi et al.	Decreases the time interval and define a high value for Imin	Improves throughput, convergence time and network lifespan	Requires more memory due to the fuzzification and defuzzification
QWL-RPL, 2020 [61]	Musaddiq et al.	Congestion status and link workload are decision parameters for objective function	Solves load unbalancing problem in diverse traffic pattern	It's very expensive to move the control messages across the network
Enhanced RPL, 2017 [62]	Hatem et al.	Integrates the value of energy and ETX with MRHOF	Throughput is increased without losing power	Doesn't well suited in random topology of network
FTC-OF, 2021 [63]	Hassani et al.	uses three metrics HC, RSSI and FTM for selecting the parent in dense network	Increase reliability, reduce power consumption and balance workload	Too much calculation require more storage space

of ignoring a valid routing pathway. For this, DIM uses Dijkstra's algorithm and calculated transmission delay and total delay. In this way, DIM guarantees real time communication in emergency response domains of IoT. The REPC takes the benefits of residual energy of SNs to select a forwarder SN in network. Hence, ERGID performs better w.r.t. end-to-end delay, packet loss and power consumption.

Due to the high implementation cost, all IoT devices aren't able to connect with internet autonomously. For that, a gateway is used for further data transmission. Sobral et al. extended the work of LOADng to achieve this, called LOADng-IoT [67].

It has three developments over LOADng w.r.t. route discovery, control messages and new error code generation. In the process of route discovery, SNs find gateway (or, internet connected device) without any prior configuration. The control message 'RREQ-IoT' is circulated to find the internet connected (IC) device. If an intermediary SN knows the route of IN, can send this control message in unicast mode. In reverse IC nodes generate RREP-IoT and send to its originator. The SN that receives RREP-IoT, create an entry on its routing set with the information of message originator. The second development is able to reduce the number of transmission, because the entries of internet route are stored inside the internet route cache (IRC). Finally, LOADng-IoT has a new error code for RERR to achieve maximum throughput.

A priority-based energy efficient (PriNergy) routing protocol is proposed by Safara et al. [68]. This is a content based routing algorithm and mainly dedicated to transmit audio and video data. Here each SN is assigned with a random digit either 0 (for high priority) or 1 (for low priority). The high priority SNs send video data while low priority SNs sends only audio clips. If there is congestion, sends only smaller amount of data i.e. audio and text. PriNergy also uses timing pattern (i.e. TDMA schedule) for data transfer from source to destination. If any error occurs in parent node, child SNs remains active until the convergence and configuration elapsed. Hence, PriNergy enhances the robustness and decrease the congestion in IoT.

Due to huge amount of data transmission in dense network area, RPL changes its path frequently from the source to destination. To solve this issue, authors introduced Energy and Data Aware Data Aggregation RPL (EDADA-RPL) [69]. It restricts the transmission of redundant data to the DODAG root of RPL network. EDADA-RPL utilizes residual energy of SNs to select the preferred parent in LLNs. Then rank of SN i.e., R(SN) is calculated by adding the value of rank of parent (R(P)) with Rank_Increase. Rank_Increase is the summation of residual energy of SNs and Min_Hop_Increase. Here, the default value of Min_Hop_Increase is 256. Compressed sensing (CS) theory is adopted to aggregate the data packets from SNs and sent it to the preferred parent. Finally, it improves packet delivery ratio, route stability, lifespan of network and minimizes delay also.

Xin Niu proposed an optimization scheme to construct the DODAG of RPL in interference environment [35]. This scheme is aiming to reduce the system power consumption and maintenance cost of network as well. It also doesn't allow the network topology to be changed frequently due to the loss of DIO. For this, RPL topology is repaired in two ways: (1) DODAG initiates global repair operation, uses incremented DODAG version_number, traverse all SNs and proposed a new DODAG version_number and (2) DODAG applies additional parameters to modernize local topology of RPL. Additionally, neighbor unreachable detection (NUD) method tells about the reachability of the SNs in the network. The NUD is explained in RFC 4861 [68] (Table 5).

Table 5 Energy efficient and QoS based routing protocols in LLNs

Routing protocol	Author's name	Methodology	Features	Limitations
Energy-IoT, 2017 [64]	Li et al.	Explores a hierarchical approach along with different duty cycles	Maximize network's life span and achieve end-to-end delay	Doesn't support heterogeneous environment of IoT
MLIA, 2019 [65]	Valls et al.	Proposes resource allocation framework for routing policy and perform data analytics tasks	Improve lifespan of network with data analytics job	Authors consider only two parameters to improve the performance of IoT
ER-GID, 2016 [61]	Qiu et al.	Introduced two methods i.e. DIM and REPC to improve the problem of ignoring a valid pathway	Reduce end-to-end delay and maintain load balancing among SNs	Number of control messages are increased and throughput is decreased
LOADng-IoT, 2019 [67]	Sobral et al.	introduce internet route discovery and cache mechanism along with a error code message for RERR	Achieve maximum throughput along with QoS in heterogeneous IoT	Takes more space in memory due to cache mechanism
PriNergy, 2020 [68]	Safara et al.	uses a content based routing approach, where the priority is assigned to each SN along with a timing pattern	Enhance robustness along with diminishing network congestion	Difficult to handle the transferring of heterogeneous data packets
EDADA-RPL, 2019 [69]	Sennan et al.	Utilizes residual energy of SNs for parent selection and further perform data aggregation	Improve packet delivery ratio along with network's lifespan	Takes longer time to envisage the parent SN
DODAG optimization, 2021 [35]	Niu et al.	Proposed an improved DODAG construction optimization scheme	reduces the system power consumption and maintenance cost in interference environment	Doesn't consider the load balancing among SNs

5.6 Security Aware Deep Learning Based RPL Routing Protocols

Dense network of IoT nodes generates huge amount of heterogeneous data leads to congestion and collision as well. To handle these issues, authors utilized Q-learning to sense the collision probability from MAC layer and makes an intelligent decision at network layer [70]. It also employs the ranking state of SNs for generating the reward function.

There is an algorithm to detect anomaly in IoT network in [71]. It is called Green Energy Efficient Routing with Deep Learning based Anomaly Detection (GEER-DLAD). It uses error lossy compression (ELC) and moth flame swarm optimization (MSO) to restrict the amount of data communicated over network and selecting the optimal path respectively. At last, anomaly is detect using recurrent neural network-long short term memory.

As Deep Learning first processes the data, paper [72] develop the dataset from RPL under Cooja Simulator. This dataset simply called IoTR-DS, which further classified as DIS, Rank and Warmhole security attacks. It first train the data using normal traffic and then labled traffic. It is a hybrid approach which combines supervised DANN and semi-supervised DAE.

Restricted Boltzmann Machine (RBM) and Long Short-Term Memory (LSTM) deep learning techniques are combined together to find the attacks on RPL network layer [73]. Total 11 parameters are chosen to predick the behavior of network (Table 6).

5.7 The Persisting Challenges

In spite of availability of a lot of options improving upon basic routing protocols, there are still requirement for performance enhancement especially in terms of energy efficiency and performance including secure communications brought in recently. Firstly, RPL was widely adopted as a routing protocol for lossy networks by IETF RoLL WG. Still RPL couldn't provide the excellent performance in different IoT scenarios and needs improvement. After discussion of existing routing protocols over RPL, we were able to identify the major constraints for current routing strategy. Total numbers of available SNs, parent SN, energy model etc. are the major input parameters, which are taken for designing an efficient routing solution. Some authors proposed routing solutions based on the RPL and other solutions of WSNs. These approaches try to solve the issues of RPL. Most of the studied routing protocols consider four parameters viz. throughput (i.e., successful transmission of data packets to the base station), stability period, power consumption and numbers of hop count to enhance the network layer performances.

To support P2P communication, IETF formulated P2P-RPL and AODV-RPL, motivated by the reactive approach of WSNs. It Increases the number of control

Table 6 Security aware deep learning based routing protocols in LLNs

Routing protocol	Author's name	Methodology	Features	Limitations
RL-based optimized RPL, 2021 [70]	Musaddiq et al.	Reinforcement based Q-Learning approach to handle congestion and collision	Lower packet ratio, improves throughput, low overhead and achieve less energy	Required to optimize the performances w.r.t trickle timer algorithm
GEER-DLAD, 2021 [71]	Lydia et al.	Uses recurrent neural network-long short term detection model to check security	Efficient utilization of power sources increases the network lifespan	Computationally complex in nature
Hybrid IDS-DL RPL, 2023 [72]	Sawafi et al.	Detect and classify cyber-attacks for IoTR-DS dataset	Detect the newly arrived attack as traffic behavior changed	Routing attacks for network layer need to be considered
DNN-RPL, 2020 [74]	Kharche et al.	Reducing interferences by deepnet layer	Minimize power consumption and delay	None
Deep RPL SDN, 2023 [75]	S Qamar	Uses reinforcement learning to predict energy level of parent SN	Minimize packet loss by 236 and energy consumption by 6%	Very few parameters are considered
RBM and LSTM based RPL, 2024 [73]	Sahay et al.	RBM and LSTM are utilized for network analysis	Detect anomaly at network layer	Need to apply hyper-parameter to tune the model

messages, which leads to more power consumption and poor network stability. Furthermore, one geographical based approach is given to reduce the control messages and power consumption in [32]. But it requires some location aware SNs (i.e., SNs equipped with GPS), which are very expensive. ER-RPL also reduces the routing overhead, but adds one more control message in the network.

Mobility is the other imperative feature, where SNs are able to move across the network for sensing the data. It is especially benefitted in case of frequent changes in network topology. Current IoT scenarios are must using the mobile sensor nodes. Fundamental RPL doesn't admit this feature. An improvement is done over RPL in this respect i.e., CO-RPL. It follows only static link and couldn't work with dynamic environment of IoT. MOD-RPL, mRPL and EKF-MRPL flood the routing table with lots of redundant control messages.

Objective function of RPL helps to choose a parent or forwarder for data transmission. OF0 and MRHOF are adopted in RPL in this direction. The OFs that consider a higher number of routing metrics tend to have better results. However, the implementation of a composed metric may not be trivial. Some solutions use fuzzy systems to perform the composition of routing metrics as an alternative to the conventional lexical and additive methods.

Thus, varieties of IoT applications seek an efficient routing algorithm for low power and lossy networks of IoT. To achieve this, proper route maintenance, parent node selection parameters and appropriate use of trickle timers are required. Considering the heterogeneous nature of IoT devices is also needed here. Because IoT devices generate various types of data like text, audio and video etc.

6 Conclusion

This paper is aiming to provide a comprehensive survey on routing protocol for low power and lossy IoT network. RPL is the most prominent routing to enable IPv6 addressing over internet. It also highlights the various design constraints and routing approach for LLNs. Thereafter review of routing algorithms are performed by considering their nature such as mobility, traffic pattern, objective function, energy efficiency and QoS etc. Along with that, review of routing protocols are provided in chronological way. No one routing protocol is there, which fulfill all routing requirement. Lots of enhancements have been done with RPL to alleviate the existing routing challenges. The comprehensive analysis and discussion of network layer routing protocols allow to identify the evolution of state-of-the-art solutions. Limitations of each routing algorithm reveals that, still they aren't fitted in low power and lossy IoT network.

References

1. Srinidhi, N., Kumar, S.D., Venugopal, K.: Network optimizations in the internet of things: a review. Eng. Sci. Technol. Int. J. **22**(1), 1–21 (2019)
2. Cruz, M.A., Rodrigues, J., Al-Muhtadi, J., Korotaev, V.V., Albuquerque, V.: A reference model for internet of things middleware. IEEE Internet Things J. **5**(2), 871–883 (2018)
3. Asghari, P., Rahmani, A.M., Javadi, H.: Internet of things applications: a systematic review. Comput. Netw. **148**, 241–261 (2019)
4. Triantafyllou, A., Sarigiannidis, P., Lagkas, T.D.: Network protocols, schemes, and mechanisms for internet of things (IoT): features, open challenges, and trends. Wirel. Commun. Mob. Comput. **2018**(1), 5349894 (2018)
5. Sohail, M., Latif, Z., Javed, S., Biswas, S., Ajmal, S., Iqbal, U., Raza, M., et al.: Routing protocols in vehicular adhoc networks (VANETs): a comprehensive survey. Internet Things **23**, 100837 (2023)
6. Liu, X., Sheng, Z., Yin, C., Ali, F., Roggen, D.: Performance analysis of routing protocol for low power and lossy networks (RPL) in large scale networks. IEEE Internet Things J. **4**(6), 2172–2185 (2017)
7. Sobral, J.V., Rodrigues, J.J., Rabêlo, R.A., Al-Muhtadi, J., Korotaev, V.: Routing protocols for low power and lossy networks in internet of things applications. Sensors **19**(9), 2144 (2019)

8. Safaei, B., Monazzah, A.M.H., Bafroei, M.B., Ejlali, A.: Reliability side-effects in internet of things application layer protocols. In: 2017 2nd International Conference on System Reliability and Safety (ICSRS), pp. 207–212. IEEE (2017)
9. Zikria, Y.B., Afzal, M.K., Ishmanov, F., Kim, S.W., Yu, H.: A survey on routing protocols supported by the Contiki internet of things operating system. Futur. Gener. Comput. Syst. **82**, 200–219 (2018)
10. Mahmud, M.A., Abdelgawad, A., Yelamarthi, K.: Energy efficient routing for internet of things (IoT) applications. In: 2017 IEEE International Conference on Electro Information Technology (EIT), pp. 442–446. IEEE (2017)
11. Aljarrah, E., Yassein, M.B., Aljawarneh, S.: Routing protocol of low-power and lossy network: survey and open issues. In: 2016 International Conference on Engineering & MIS (ICEMIS), pp. 1–6. IEEE (2016)
12. Clausen, T., Yi, J., Herberg, U.: Lightweight on-demand ad hoc distance-vector routing-next generation (LOADng): protocol, extension, and applicability. Comput. Netw. **126**, 125–140 (2017)
13. Kharrufa, H., Al-Kashoash, H.A., Kemp, A.H.: RPL-based routing protocols in IoT applications: a review. IEEE Sens. J. **19**(15), 5952–5967 (2019)
14. Vermesan, O., Friess, P.: Internet of Things Applications-from Research and Innovation to Market Deployment. Taylor & Francis (2014)
15. Li, Q., Gochhayat, S.P., Conti, M., Liu, F.: EnergIoT: a solution to improve network lifetime of IoT devices. Pervasive Mob. Comput. **42**, 124–133 (2017)
16. Kushalnagar, N., Montenegro, G., Schumacher, C.: IPv6 over low-power wireless personal area networks (6LoWPANs): overview, assumptions, problem statement, and goals (2007)
17. Ko, J., Terzis, A., Dawson-Haggerty, S., Culler, D.E., Hui, J.W., Levis, P.: Connecting low-power and lossy networks to the internet. IEEE Commun. Mag. **49**(4), 96–101 (2011)
18. Schumacher, C.: IPv6 over low-power wireless personal area networks (6LoWPANs): overview, assumptions, problem statement, and goals status of this memo this memo provides information for the internet community. It does not specify an internet standard of any kind. Distribution of this (2007)
19. Montenegro, G.: Transmission of IPv6 packets over IEEE 802.15. 4 networks. Technical report, No. RFC 4944 (2007)
20. Hui, J., Thubert, P.: Compression format for IPv6 datagrams over IEEE 802.15. 4-based networks. Technical report (2011)
21. Kim, E., Kaspar, D., Vasseur, J.: Design and application spaces for IPv6 over low-power wireless personal area networks (6LoWPANs). Technical report (2012)
22. Kim, E., Kaspar, D., Gomez, C., Bormann, C.: Problem statement and requirements for IPv6 over low-power wireless personal area network (6LoWPAN) routing. Technical report (2012)
23. Shelby, Z., Chakrabarti, S., Nordmark, E., Bormann, C.: Neighbor discovery optimization for IPv6 over low-power wireless personal area networks (6LoWPANs). Technical report (2012)
24. Brandt, A., Hui, J., Kelsey, R., Levis, P., Pister, K., Struik, R., Vasseur, J., Alexander, R.: RPL: IPv6 routing protocol for low-power and lossy networks. In: RFC 6550 (2012)
25. Sheng, Z., Yang, S., Yu, Y., Vasilakos, A.V., McCann, J.A., Leung, K.K.: A survey on the IETF protocol suite for the internet of things: standards, challenges, and opportunities. IEEE Wirel. Commun. **20**(6), 91–98 (2013)
26. Dohler, M., Watteyne, T., Winter, T., Barthel, D.: Routing requirements for urban low-power and lossy networks. Technical report (2009)
27. Pister, K., Thubert, P., Dwars, S., Phinney, T.: Industrial routing requirements in low-power and lossy networks. Technical report (2009)
28. Brandt, A., Buron, J., Porcu, G.: Home automation routing requirements in low-power and lossy networks. Technical report (2010)
29. Martocci, J., De Mil, P., Riou, N., Vermeylen, W.: Building automation routing requirements in low-power and lossy networks. Technical report (2010)
30. Singh, D., Maurya, A.K., Dewang, R.K., Keshari, N.: A review on internet of multimedia things (IoMT) routing protocols and quality of service. In: Internet of Multimedia Things (IoMT), pp. 1–29 (2022)

31. Poornima, M., Vimala, H., Shreyas, J.: Holistic survey on energy aware routing techniques for IoT applications. J. Netw. Comput. Appl. **213**, 103584 (2023)
32. Musaddiq, A., Zikria, Y.B., Zulqarnain, Kim, S.W.: Routing protocol for low-power and lossy networks for heterogeneous traffic network. EURASIP J. Wirel. Commun. Netw. **2020**, 1–23 (2020)
33. Altan, G.: SecureDeepNet-IoT: a deep learning application for invasion detection in industrial internet of things sensing systems. Trans. Emerg. Telecommun. Technol. **32**(4), 4228 (2021)
34. Jayabalan, E., Pugazendi, R.: Deep learning model-based detection of jamming attacks in low-power and lossy wireless networks. Soft. Comput. **26**(23), 12893–12914 (2022)
35. Niu, X.: [Retracted] optimizing DODAG build with RPL protocol. Math. Probl. Eng. **2021**(1), 5579564 (2021)
36. Levis, P., Clausen, T., Hui, J., Gnawali, O., Ko, J.: RFC 6206: the trickle algorithm. RFC Editor (2011)
37. Thubert, P.: Objective function zero for the routing protocol for low-power and lossy networks (RPL). Technical report (2012)
38. Gnawali, O., Levis, P.: The minimum rank with hysteresis objective function. Technical report (2012)
39. Lamaazi, H., Benamar, N., Jara, A.J.: RPL-based networks in static and mobile environment: a performance assessment analysis. J. King Saud Univ. Comput. Inf. Sci. **30**(3), 320–333 (2018)
40. Manrique, J.A., Rueda-Rueda, J.S., Portocarrero, J.M.: Contrasting internet of things and wireless sensor network from a conceptual overview. In: 2016 IEEE International Conference on Internet of Things (iThings) and IEEE Green Computing and Communications (GreenCom) and IEEE Cyber, Physical and Social Computing (CPSCom) and IEEE Smart Data (SmartData), pp. 252–257. IEEE (2016)
41. Goyal, M., Baccelli, E., Philipp, M., Brandt, A., Martocci, J.: Reactive discovery of point-to-point routes in low-power and lossy networks. Technical report (2013)
42. Zhao, M., Kumar, A., Joo Chong, P.H., Lu, R.: A comprehensive study of RPL and P2P-RPL routing protocols: implementation, challenges and opportunities. Peer-to-Peer Netw. Appl. **10**, 1232–1256 (2017)
43. Barriquello, C.H., Denardin, G.W., Campos, A.: A geographic routing approach for IPv6 in large-scale low-power and lossy networks. Comput. Electr. Eng. **45**, 182–191 (2015)
44. Zhao, M., Ho, I.W.-H., Chong, P.: An energy-efficient region-based RPL routing protocol for low-power and lossy networks. IEEE Internet Things J. **3**(6), 1319–1333 (2016)
45. Anamalamudi, S., Zhang, M., Sangi, A.R., Perkins, C.E., Anand, S.: BL asymmetric AODV-P2P-RPL in low-power and lossy networks (LLNs). Technical report, Internet-Draft draft-ietf-roll-aodv-rpl-03; Work in Progress
46. Khan, M.M., Lodhi, M.A., Rehman, A., Khan, A., Hussain, F.B.: Sink-to-sink coordination framework using RPL: routing protocol for low power and lossy networks. J. Sens. **2016**(1), 2635429 (2016)
47. Gaddour, O., Koubäa, A., Rangarajan, R., Cheikhrouhou, O., Tovar, E., Abid, M.: Co-RPL: RPL routing for mobile low power wireless sensor networks using corona mechanism. In: Proceedings of the 9th IEEE International Symposium on Industrial Embedded Systems (SIES 2014), pp. 200–209. IEEE (2014)
48. Fotouhi, H., Moreira, D., Alves, M.: mRPL: boosting mobility in the internet of things. Ad Hoc Netw. **26**, 17–35 (2015)
49. Gara, F., Saad, L.B., Ayed, R.B., Tourancheau, B.: RPL protocol adapted for healthcare and medical applications. In: 2015 International Wireless Communications and Mobile Computing Conference (IWCMC), pp. 690–695. IEEE (2015)
50. Fotouhi, H., Moreira, D., Alves, M., Yomsi, P.M.: mRPL+: a mobility management framework in RPL/6LoWPAN. Comput. Commun. **104**, 34–54 (2017)
51. Bouaziz, M., Rachedi, A., Belghith, A.: EKF-MRPL: advanced mobility support routing protocol for internet of mobile things: movement prediction approach. Futur. Gener. Comput. Syst. **93**, 822–832 (2019)

52. Yi, J., Clausen, T.: Collection tree extension of reactive routing protocol for low-power and lossy networks. Int. J. Distrib. Sens. Netw. **10**(3), 352421 (2014)
53. Kim, H.-S., Cho, H., Kim, H., Bahk, S.: DT-RPL: diverse bidirectional traffic delivery through RPL routing protocol in low power and lossy networks. Comput. Netw. **126**, 150–161 (2017)
54. Taghizadeh, S., Bobarshad, H., Elbiaze, H.: CLRPL: context-aware and load balancing RPL for IoT networks under heavy and highly dynamic load. IEEE Access **6**, 23277–23291 (2018)
55. Lee, S., Jeong, Y., Moon, E., Kim, D.: An efficient mop decision method using hop interval for RPL-based underwater sensor networks. Wirel. Pers. Commun. **93**, 1027–1041 (2017)
56. Ko, J., Jeong, J., Park, J., Jun, J.A., Gnawali, O., Paek, J.: DualMOP-RPL: supporting multiple modes of downward routing in a single RPL network. ACM Trans. Sens. Netw. (TOSN) **11**(2), 1–20 (2015)
57. Sousa, N., Sobral, J.V., Rodrigues, J.J., Rabêlo, R.A., Solic, P.: ERAOF: a new RPL protocol objective function for internet of things applications. In: 2017 2nd International Multidisciplinary Conference on Computer and Energy Science (SpliTech), pp. 1–5. IEEE (2017)
58. Bouzebiba, H., Lehsaini, M.: FreeBW-RPL: a new RPL protocol objective function for internet of multimedia things. Wirel. Pers. Commun. **112**(2), 1003–1023 (2020)
59. Lamaazi, H., Benamar, N.: RPL enhancement using a new objective function based on combined metrics. In: 2017 13th International Wireless Communications and Mobile Computing Conference (IWCMC), pp. 1459–1464. IEEE (2017)
60. Lamaazi, H., Benamar, N.: A novel approach for RPL assessment based on the objective function and trickle optimizations. Wirel. Commun. Mob. Comput. **2019**(1), 4605095 (2019)
61. Musaddiq, A., Zikria, Y.B., Zulqarnain, Kim, S.W.: Routing protocol for low-power and lossy networks for heterogeneous traffic network. EURASIP J. Wirel. Commun. Netw. **2020**, 1–23 (2020)
62. Hatem, J.A., Safa, H., El-Hajj, W.: Enhancing routing protocol for low power and lossy networks. In: 2017 13th International Wireless Communications and Mobile Computing Conference (IWCMC), pp. 753–758. IEEE (2017)
63. Hassani, A.E., Sahel, A., Badri, A.: FTC-OF: forwarding traffic consciousness objective function for RPL routing protocol. Int. J. Electr. Electron. Eng. Telecommun. **10**(3), 168–175 (2021)
64. Li, Q., Gochhayat, S.P., Conti, M., Liu, F.: EnergIoT: a solution to improve network lifetime of IoT devices. Pervasive Mob. Comput. **42**, 124–133 (2017)
65. Valls, V., Iosifidis, G., Salonidis, T.: Maximum lifetime analytics in IoT networks. In: IEEE INFOCOM 2019-IEEE Conference on Computer Communications, pp. 1369–1377. IEEE (2019)
66. Qiu, T., Lv, Y., Xia, F., Chen, N., Wan, J., Tolba, A.: ERGID: an efficient routing protocol for emergency response internet of things. J. Netw. Comput. Appl. **72**, 104–112 (2016)
67. Sobral, J.V., Rodrigues, J.J., Rabêlo, R.A., Saleem, K., Furtado, V.: LOADng-IoT: an enhanced routing protocol for internet of things applications over low power networks. Sensors **19**(1), 150 (2019)
68. Safara, F., Souri, A., Baker, T., Al Ridhawi, I., Aloqaily, M.: PriNergy: a priority-based energy-efficient routing method for IoT systems. J. Supercomput. **76**(11), 8609–8626 (2020)
69. Sennan, S., Balasubramaniyam, S., Luhach, A.K., Ramasubbareddy, S., Chilamkurti, N., Nam, Y.: Energy and delay aware data aggregation in routing protocol for internet of things. Sensors **19**(24), 5486 (2019)
70. Musaddiq, A., Ali, R., Choi, J.-G., Kim, B.-S., Kim, S.W.: Collision observation-based optimization of low-power and lossy IoT network using reinforcement learning. Comput. Mater. Cont. **67**(1), 799–814 (2021)
71. Lydia, E.L., Jovith, A.A., Devaraj, A., Seo, C., Joshi, G.P.: Green energy efficient routing with deep learning based anomaly detection for internet of things (IoT) communications. Mathematics **9**(5), 500 (2021)
72. Al Sawafi, Y., Touzene, A., Hedjam, R.: Hybrid deep learning-based intrusion detection system for RPL IoT networks. J. Sens. Actuator Netw. **12**(2), 21 (2023)

73. Sahay, R., Nayyar, A., Shrivastava, R.K., Bilal, M., Singh, S.P., Pack, S.: Routing attack induced anomaly detection in IoT network using RBM-LSTM. ICT Express (2024)
74. Kharche, S., Pawar, S.: Optimizing network lifetime and QoS in 6LoWPANs using deep neural networks. Comput. Electr. Eng. **87**, 106775 (2020)
75. Qamar, S.: Optimal sensor network routing with secure network monitoring using deep learning architectures. Neural Comput. Appl. **35**(26), 19039–19050 (2023)

Generative Artificial Intelligence Using Deep Learning on Wireless Ad-Hoc Networks

Antonio Cortés

Abstract In this article, we discuss the use of Generative Artificial Intelligence (GenAI) to improve the efficiency and performance of access to wireless points located in various spaces and specific places, which allows interaction with wireless mesh networks and allows the use of mobile devices to access to everything types of information in the internal environment. Furthermore, we propose the use of generative neural networks, which are one of the pillars of the GenAI, since they use a methodology from the perspective of Deep Learning, which allows analyzing a large amount of data and detecting certain types of patterns that help a better placement of access points for better reception and connectivity. Images (heat maps), access point locations, positioning points, and bandwidth are analyzed, allowing new information to be created. On the other hand, to understand and model the general architecture of the wireless Ad-Hoc network, we use two processes that are part of neural networks, such as Multilayer Perceptron (MPL), and the Radial Basis Function (RBF), which it is a function of predictors or independent variables or input variables, which allows the prediction error in the object or output variables of the wireless network architecture to be reduced. Using these two processes does help reduce blind spots, in those internal places where the wireless signal does not reach, resulting in a signal drop. Improving internal scenarios with wireless Ad-Hoc networks is what is required for better functioning and performance of the network infrastructure.

Keywords Generative artificial intelligence (GenAI) · Deep learning · Wireless Ad-Hoc network · Mesh network · Multilayer perceptron (MPL) · Radial basis function (RBF)

A. Cortés (✉)
Departamento de Informática, Universidad de Panamá, Panama City, Panama
e-mail: antonio.cortes@up.ac.pa

G. ALTAN and I. ABASIKELEŞ TURGUT (eds.), *Deep Learning in Ad-Hoc Wireless Networks*, Studies in Big Data 172,
https://doi.org/10.1007/978-3-031-86075-1_4

1 Introduction

It has been observed that in recent years there has been an evolution in wireless communication networks, which allows the number of nodes for wireless networks to grow exponentially. At the same time, with the emergence and integration of the Internet of Things (IoT) and the incorporation of new technologies such as Cloud Computing, Edge Computing and Software Defined Networks [1], it has been possible to integrate variety of services and applications in wireless networks, which means that Internet Service Providers (ISP) have to improve their network infrastructures to provide greater bandwidth in the shortest possible time due to the massive consumption of data in real time.

Likewise, having this large number of access points or active nodes, in which you have a variety of mobile users moving in all directions in small and specific spaces, accessing all types of valuable and diversified information, in most cases. Neverless, through mobile devices, laptops or digital tablets, security [2] when accessing these nodes is of very high relevance, because collisions at the frequency level between these nodes can sometimes occur, which weakens the bandwidth and access to data, generating blind or dead spots, where the wireless signal does not reach and therefore allows unwanted intruders to access the wireless network, generating some type of fraud or infecting the network with a worm, trojan, rumors, etc.

Given this situation that arises with access to data in wireless networks, we have given ourselves the task of using Generative Artificial Intelligence (GenAI), which is a subbranch of Artificial Intelligence that allows, from existing data that is generated through a series of parameters in wireless network infrastructures generate new original data and analysis of these that help to improve performance and security in wireless networks.

In this research article we set ourselves the following objectives:

- Analyze how wireless Ad-Hoc network infrastructures improve their performance, become more efficient and produce better benefits to users using Generative Artificial Intelligence using a Deep Learning approach taking Generative Neural Networks as reference.
- Collect those predictors to later parameterize them, such as signal levels, signal-tonoise ratio, signal-to-interference ratio, number of access points, noise levels, frequency band coverage, network coverage PHY mode of the 802.11 protocols (a, b, g, n, ac, ax), which allow the neural network to be modeled from the processes called Multilayer Perceptron (MLP) and the Radial Basis Function (RBF).
- Model the structure of the wireless Ad-Hoc network from Deep Learning, taking into consideration the independent or input variables which will be represented through Generative Neural Networks visualized through MPL and RBF processes, respectively.

  In this same direction, it is important to point out the importance of the parametrization of the independent variables, as well as the normalization of these, since it helps to create a model with predictive and residual value from the entry of these covariates or number of units. At the same time, in Fig. 1, we can have several

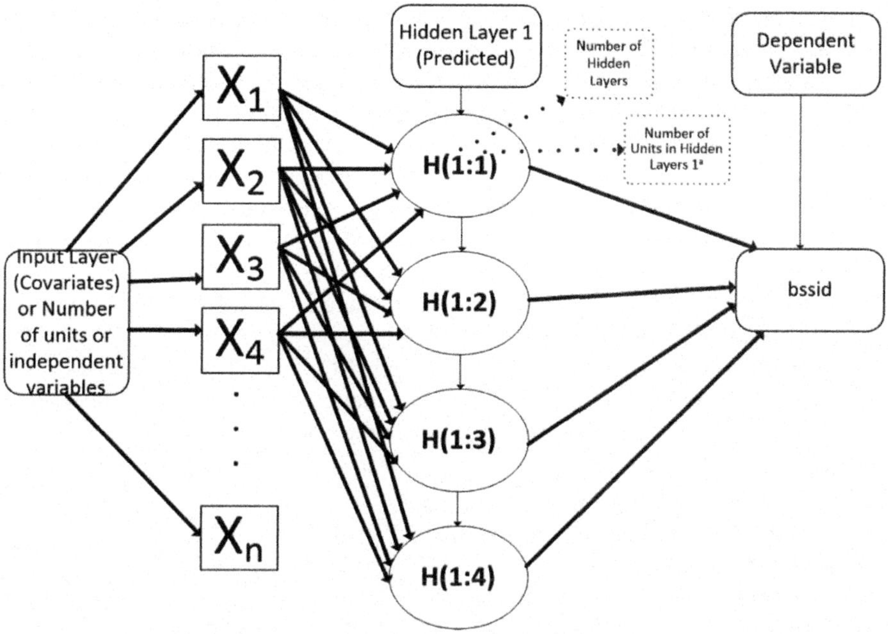

Fig. 1 Model with prediction value

hidden layers and for each of these layers several units in these hidden layers that help the performance of the model by activating various functions, for example, the hyperbolic tangent, among others.

Indistinctly, regardless of the model that we use for the wireless network infrastructure from the use of the GenAI and making use of the two processes proposed in the neural network, MLP and RBF, respectively, it should allow us from the normalized independent variables the selection of predictive values that allow the designed wireless network infrastructure to improving.

In turn, highly innovative generative machine artificial intelligence (GenAI) models powered by deep learning and neural networks can enable the digital transformation of organizations, thereby increasing productivity, efficiency, and problem-solving capabilities. Then, based on these predicted values, original content is generated to address complex challenges, allowing GenAI to revolutionize not just how organizations operate, but how innovation is structured.

However, if we take into consideration the increased use of Deep Learning and GenAI level applications with large data volume and computational complexity, this allows unprecedented demands on wireless infrastructures, requiring reliable, high-bandwidth, low-latency data, and significantly higher structured cabling densities, and advanced cooling methods.

So far, wireless network infrastructure is preparing for artificial intelligence, and users need powerful and innovative network infrastructure solutions to help with the design, implementation, and scalability of back-end, front-end, and ambient network architecture storage for complex high-performance computing (HPC).

We also resort to the accelerator models of GenIA and Deep Learning, which consist of training (where a new aspect is learned) and inference (applying what is learned to new data). Typically, these neural and Deep Learning networks mimic the architecture and functions of the human brain to learn and generate new original knowledge by considering the analysis of patterns, nuances, and the general and variable characteristics of a massive and complex data set. Likewise, large language models (LLM), such as ChatGPT and Google Bard, are clear examples of these GenAI models, which are trained from large amounts of data to understand and generate plausible linguistic responses.

Ultimately, general-purpose CPUs that perform I/O and control operations sequentially cannot effectively extract large volumes of data in parallel from multiple sources and processes quickly enough, so Deep Accelerated Learning and GenAI use parallel processing-based graphics processing units (GPUs) to execute thousands of high-performance calculations.

Following the introduction, this article is divided into the following sections: Sect. 2 reviews related work in brief. In Sect. 3 we describe our proposed architecture. Section 4 reports our experimental results and analysis. Section 5 Discussion. Conclusion and future work of this paper will be presented in Sect. 6.

2 Related Works

In [3], having signaled from several cells implies having frequency interference between different channels, which causes calls made through one of these channels to be distorted. Therefore, the authors propose to eliminate this noise in the channel to use a multilayer perceptron network trained with Wilcoxon learning, which will help stabilize the inputs and outputs produced in the system through the wireless network system.

On the other hand, in [4] the authors use the multi-layer perceptron network applied to Wireless Sensory Networks to analyze security in this type of networks. This is used to analyze the CSMA protocol located at the physical level, specifically at the MAC layer to analyze the various denial of service attacks launched by unethical hackers. Likewise, in [5] the authors propose using a model based on multilayer perceptron-type networks to be trained from a set of data, in such a way that it helps them optimize network congestion and use of the bandwidth. The network type they use is wireless mesh backbone networks.

Therefore, in [6] the authors use artificial neural networks by making use of the Multilayer Perceptron for the detection and classification of a diversity of attacks on wireless Ad-Hoc networks. In turn, in [7], the authors propose that there are malicious nodes that can destroy data in Mobile Ad-Hoc networks, and that to predict how much damage these viruses, malware, fake news, rumors, exploits can cause, in the networks use neural networks, specifically the Multilayer Perceptron, to identify which of these perceptron are the most harmful and can cause the most damage to the network. In this same direction, in [8] possible predictions are outlined using the Multilayer Perceptron to improve mobility in wireless networks since the processes of connection and reconnection to the various access points, whether internal or external, of a certain space decreases network performance. Similarly, in [9], the authors use multilayer distributed perceptron (MLPC) to detect various types of distributed denial of service (DDoS) attacks, in modern vehicular communication systems. They use Apache Spark-level simulation processes to create the MLPC and Amazon Web Services (AWS) to train and determine the attack time at the distributed neural network level. However, in [10] the authors use a Multilayer Perceptron neural network with 4 hidden layers with 20 hidden units to detect dead links between nodes in complex networks.

3 Materials and Methods

In this section, we describe our approach to Generative Artificial Intelligence using deep learning in wireless Ad-Hoc networks. Basically, there are two main tasks: selection and classification. In Fig. 2, our proposed architecture is presented, which begins with the data collection task, then the training phase, and ends with the validation phase.

To validate our proposed architecture, we use experimental data, and this data is tested through simulation, in which active and offline access points are considered

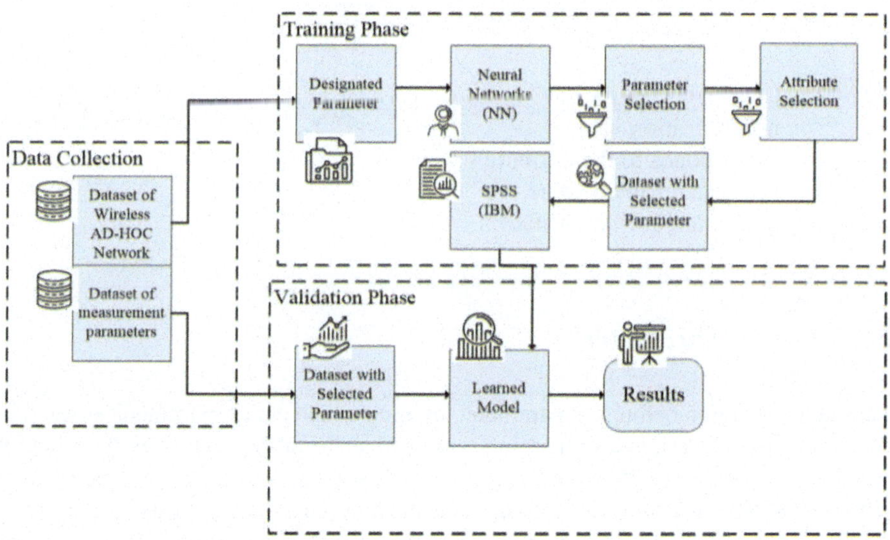

Fig. 2 Our proposed architecture

by creating data repositories of Ad-Hoc wireless networks, which store a series of parameters. The set of processed parameters is introduced into the feature selection method. The parameter selection method contained in the learning of neural networks (NN) results in a list of important parameters which are classified according to the level of importance. To validate this shortened list of parameters we use Deep Learning represented using Multilayer Perceptron (MLP) neural networks and the Radial Basis Function (RBF).

3.1 Experimental Data

The Ad-Hoc wireless network data set not only contains discrete type values, but also numerical type values, character strings, which allow generating scalar and nominal type measurements. At the same time, the processing of experimental data must be carried out in advance. Therefore, two main stages are presented for the processing with experimental data, where first we must carefully select the parameters to transform them into attributes and subsequently assign them a numerical value and second the normalization stage of these experimental data. It is worth mentioning that some variables such as security are considered as a string of characters while others such as bssid, down_speed, up_load_speed, wireless_transmit_rate, among others, are considered to have numeric values. Once all the attribute values are converted to integer values, each of the attributes is linearly normalized between zero and one. Equation (1) displays the normalization formula.

$$N_i = \frac{x_i - \min(x)}{\max(x) - \min(x)} \tag{1}$$

where:

N_i	normalization value,
X_i	corresponds to the attribute's value,
$\min(x)$	minimal attribute value,
$\max(x)$	maximum attribute value.

3.2 Setting Parameters

Parameter selection belongs to parameter learning which entails parameter extraction and selection [11]. Parameter learning is defined as the ability to model the behavior of data traffic between different access points based on raw information from more parameters. The learning that can take place using parameters is very important to show the correlation between the detection performance and the quality of the data traffic model [12].

In the same direction, parameter extraction and selection are different terms. Parameter extraction refers to obtaining new parameters from a space in which other unprocessed parameters are located so that they are informative and non-redundant. Those parameters that have not been processed yet and the newly generated parameters are usually different. On the other hand, parameter selection consists of selecting several parameters from the raw parameter space. Therefore, the generated parameters are simply selected from original parameters without transformation.

Likewise, both parameter extraction and selection aim for a smaller number of new parameters generated from the original parameters. In this article, we adopt parameter selection using neural networks (NN) explicitly, while parameter extraction is implemented through the statistical analysis tool SPSS 25, implicitly. We use NN to improve the bandwidth and load balance that is generated from the access points, in such a way that the areas or zones or dead spots where there is no wireless signal reception can be reduced. By using NN, we can choose some parameters that are important for learning from the neural network called Multilayer Perceptron (MLP) and Radial Basis Function (RBF), based on the heuristic weights of learning the NN. We trained our NN using thirteen to fourteen variables, with two hidden layers and hyperbolic target and sigmoid functions. It is important to note that the more hidden layers there are and for each of these layers, the number of units, covariates or independent variables increases, the model to be executed becomes slower when executed and even the results to be obtained take longer than usual estimated. Figure 3 shows the NN model where b1 (bias 1) and b2 (bias 2), respectively, represent the bias values for the corresponding hidden layer, specifically.

However, the first hidden layer is used for parameter selection, and we consider the weight value between the first two layers to select important input parameters. The weight represents the contribution of input features to the features in the first hidden layer. The weights of the start and end nodes represented by W_{ij}, as seen in Fig. 4, very small or even zero mean that the input parameters, corresponding to X_j, are meaningless for further propagation.

Therefore, one hidden layer is enough since we consider the weights in the first hidden layer only. We define the importance value of each input variable or unit or covariant or independent variable, as expressed in Eq. (2).

$$V_j = \sum_{i=1}^{h} |W_{ij}| \tag{2}$$

where:

h represents the number of neurons in the first hidden layer,

V_j allows selecting the most relevant parameters, ordered according to the input parameters ordered descending. We select some parameters that have a V_j value larger than a threshold value (μ).

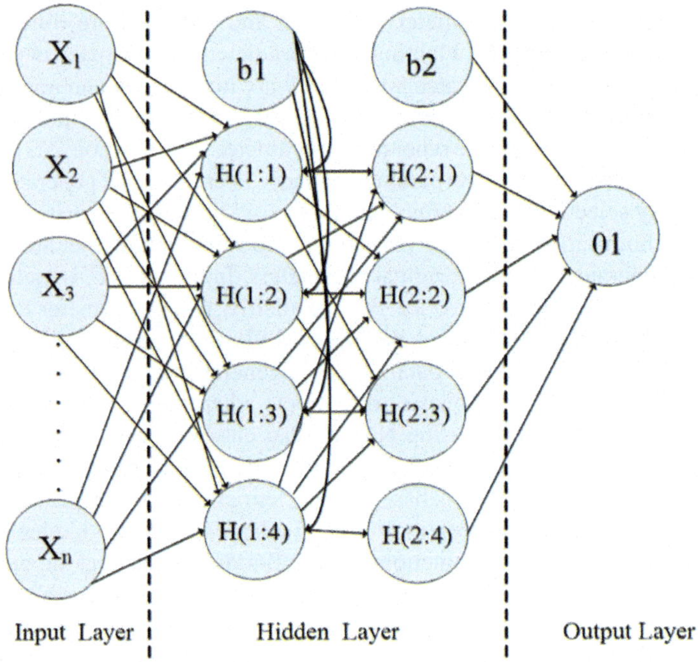

Fig. 3 Neural network (NN) model

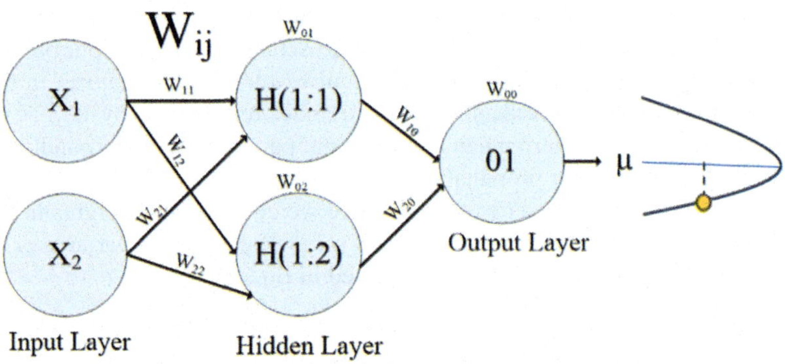

Fig. 4 Representation of the weights of the start and end nodes (W_{ij})

3.3 Parameter Classification

To validate the performance of the chosen parameters we use a deep learning algorithm. Deep learning, in turn, is a class of machine learning method that exploits cascading layers of data processing stages in a hierarchical structure for learning these supervised parameters and for pattern classification. We chose SPSS 25 as a

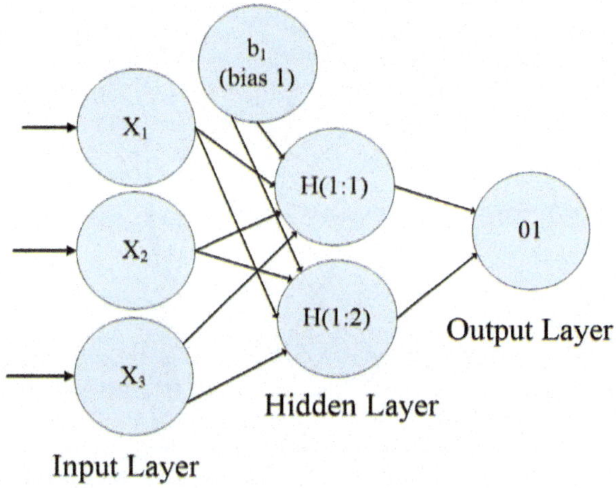

Fig. 5 MLP model

classifier because it contains a neural network called Multilayer Perceptron (MLP) that allows replacement of original parameters from a supervised approach and has a hierarchical feature extraction phase. Likewise, the MLP neural network is like an NN, as shown in Fig. 5.

Compared to the NN model, the MLP neural network is characterized by several inputs, one or two levels of hidden layers with two or four levels of the number of elements per hidden layer and a single output. Meanwhile, the nodes in the middle of the hidden layer represent a set of new parameters with high dimensions. This architecture allows data reuse after complex calculations have been executed. Likewise, the MLP neural network aims to learn from a compact set of data efficiently, so it can be agglomerated to construct deep networks. However, the results obtained from each training in the intermediate layer can be cascaded. This structure is called MLP, and it can learn from different new parameters at different depth levels. Figure 6 shows the proposed MLP architecture used in this article.

4 Experimentation

In this section we show the different stages in detail that have been carried out at the experiment level and the results obtained from them. We first explain our tests and data set preparation followed by the experimental results and analysis.

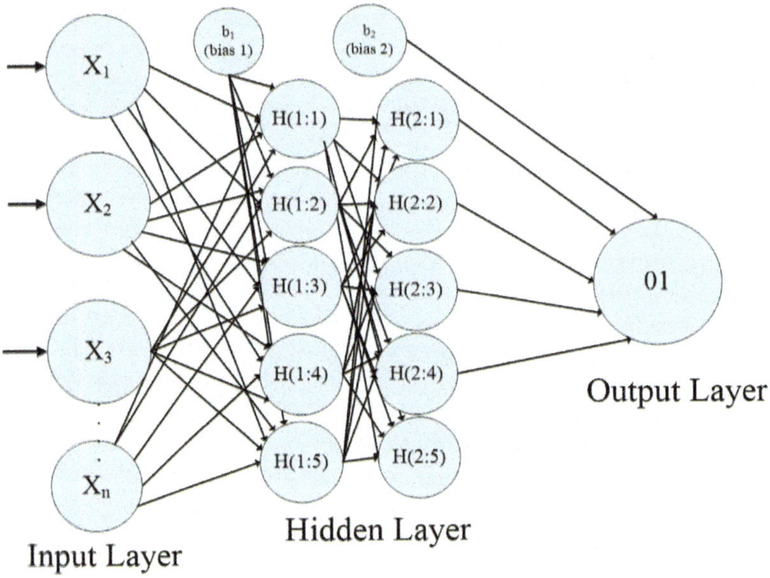

Fig. 6 Proposed MLP architecture

4.1 Experimental Setup

We constructed a data document using IBM SPSS version 25, with a total of 13 variables to demonstrate that our proposed scheme can improve not only the bandwidth and data load balance between the interconnection of different access points in an Ad-Hoc wireless network when using Generative Artificial Intelligence. Table 1 shows the summary of the parameters used for testing.

Two experimental tests are described below:

1. **Experimental Test No. 1**

In this first test, 13 variables with active access points and with a double hidden layer neural network containing 4 hidden units are used. There is the following network information, which is seen in the following Table 2.

It is important to highlight that at the level of the input layer, the rescaling method for standardized covariates or min-max normalization is presented, and the one we use is the standardized one. At the hidden layer level, we have the activation function which is used as information of the network, and we use the hyperbolic tangent. On the other hand, at the output layer level, we have the scale-dependent rescaling method and in this case we use adjusted normalization. Similarly, at the level of the activation function we use the hyperbolic tangent and as the error function the sum of the squares is used.

Table 1 Summary of parameters used in testing

ID	Parameter name	Type	Width	Decimal	Label	Columns	Align	Measure	Role
1	down_load	Numeric	3	0	Download_speed	14	Right	Nominal	Input
2	upload_speed	Numeric	3	0	upload_speed	12	Right	Nominal	Input
3	wireless_transmit_rate	Numeric	3	0	wireless_transmit_rate	16	Right	Nominal	Input
4	signal_level	Numeric	2	0	signal_level	10	Right	Nominal	Input
5	signal_to_noise_radio	Numeric	2	0	signal_to_noise_radio	17	Right	Nominal	Input
6	signal_to_interference_radio	Numeric	2	0	signal_to_interference_radio	20	Right	Nominal	Input
7	noise_level	Numeric	2	0	noise_level	10	Right	Nominal	Input
8	Issues_with_SNR	Numeric	2	0	Issues_with_SNR	14	Right	Nominal	Input
9	low_signal_level	Numeric	2	0	low_signal_level	13	Right	Nominal	Input
10	high_level_of_noise	Numeric	2	0	high_level_of_noise	14	Right	Nominal	Input
11	overlapping_channels_SIR	Numeric	2	0	overlapping_channels_SIR	19	Right	Nominal	Input
12	low_download_rate	Numeric	2	0	low_download_rate	15	Right	Nominal	Input
13	low_upload_rate	Numeric	2	0	low_upload_rate	13	Right	Nominal	Input

Table 2 Network information for experimental test No. 1

Input layer	Hidden layer(s)			Output layer	
Covariates	Number of hidden layers	Number of units in hidden layer 1[a]	Number of units in hidden layer 2[a]	Dependent variables	Number of units
13	2	4	4	1	1

In Fig. 7, you can see how the neural network is constructed from the Multilayer Perceptron (MLP).

In Fig. 8, the estimation of the parameters is observed at the neural network level.

In Fig. 8, we can observe the estimated values for the issues_with_SNR parameter in the hidden layer one and the number of elements for each of the layers. In Table 3, we can see the level of importance of the independent variables that is derived from the estimation of the parameters.

Between the signal-to-noise ratio and the overlap between the various channels covered by the wireless spectrum considerably affect the performance at the level of bandwidth and the load balance between the various access points that may be connected at a certain time.

2. **Experimental Test No. 2**

In this second test, 13 variables with active access points and with a double hidden layer neural network containing 8 hidden units are used. There is the following network information, which is seen in the following Table 4.

In this new network information, what changes with respect to experimental test No. 1 is the amount of the number of units in the first and second hidden layers, since we go from 4 to 8 units in the number of hidden layers that there are 2. In Fig. 10, you can see how the neural network is constructed from the Multilayer Perceptron (MLP).

To the extent that the number of hidden layers increases and with it the number of units for each of these layers, as we can see in Fig. 10, the level of complexity of the Multilayer Perceptron is increasingly complex, which helps obtain a more accurate and efficient predictive value.

In Fig. 9, the estimation of the parameters is observed at the neural network level. We can observe the estimated values for the signal_to_interference_ratio parameter in the hidden layer one and the number of elements for each of the layers.

In Table 5, we can see the level of importance of the independent variables that is derived from the estimation of the parameters

In Table 5, we can observe that the level of importance 0.159 which has a normalized importance of 100.0% falls on the independent variable signal_to_interference_ratio, which effectively shows us that the overlap that may occur in the waves emitted by the Access points can affect performance in terms of bandwidth and the management of large volumes of data in the wireless Ad-Hoc network. Another parameter that indicates low performance in this type of networks is related to the parameter

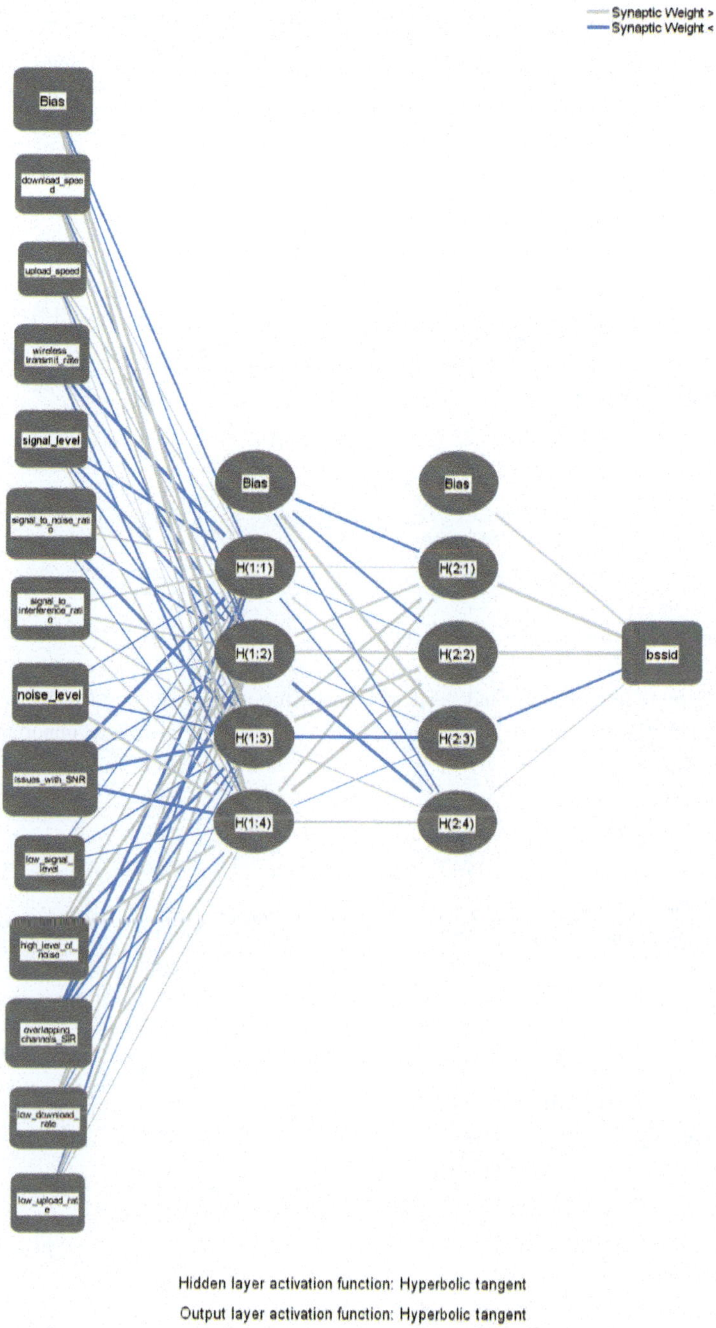

Fig. 7 Neural network based on MLP for 2 hidden layers and 4 units for the 1st and 2nd hidden layers for experimental test No. 1

Independent Variable Importance

	Importance	Normalized Importance
download_speed	.059	36.1%
upload_speed	.023	14.0%
wireless_transmit_rate	.059	35.9%
signal_level	.047	28.9%
signal_to_noise_ratio	.137	83.6%
signal_to_interference_ratio	.082	50.1%
noise_level	.066	40.3%
issues_with_SNR	.164	100.0%
low_signal_level	.034	21.0%
high_level_of_noise	.082	49.9%
overlapping_channels_SIR	.120	73.2%
low_download_rate	.072	44.1%
low_upload_rate	.054	32.9%

Fig. 8 Parameter estimation for experimental test No. 1

Table 3 Level of importance of independent variables for experimental test No. 1

Independent variables	Importance	Normalized importance (%)
download_speed	0.059	36.1
upload_speed	0.023	14.0
wireless_transmit_rate	0.059	35.9
signal_level	0.047	28.9
signal_to_noise_ratio	0.137	83.6
signal_to_interference_ratio	0.082	50.1
noise_level	0.066	40.3
issues_with_SNR	0.164	100.0
low_signal_level	0.034	21.0
high_level_of_noise	0.082	49.9
overlapping_channels_SIR	0.120	73.2
low_download_rate	0.072	44.1
low_upload_rate	0.054	32.9

Table 4 Network information for experimental test No. 2

Input layer	Hidden layer(s)			Output layer	
Covariates	Number of hidden layers	Number of units in hidden layer 1[a]	Number of units in hidden layer 2[a]	Dependent variables	Number of units
13	2	8	8	1	1

Independent Variable Importance

	Importance	Normalized Importance
download_speed	.028	17.6%
upload_speed	.079	49.6%
wireless_transmit_rate	.037	23.4%
signal_level	.064	40.1%
signal_to_noise_ratio	.039	24.8%
signal_to_interference_ratio	.159	100.0%
noise_level	.038	23.6%
issues_with_SNR	.088	55.1%
low_signal_level	.138	86.8%
high_level_of_noise	.110	68.9%
overlapping_channels_SIR	.077	48.2%
low_download_rate	.072	45.0%
low_upload_rate	.071	44.2%

Fig. 9 Parameter estimation for experimental test No. 2

called low_signal_level, which has an importance level of 0.138 and a normalized importance of 86.8%, since this is due precisely to the collision at the level waves that the access points may present.

4.2 Dataset

The grouping of the independent variables rests in the data repository that we have called SSID-GWi-Fi-Hot Access Points.sav, which has been built and designed under the standards of the SPSS statistical tool version 25 of IBM. Each of the covariates is assigned an identifier with its respective name, data type, field width, decimals and a label with which each of these independent variables is identified, a value field, the

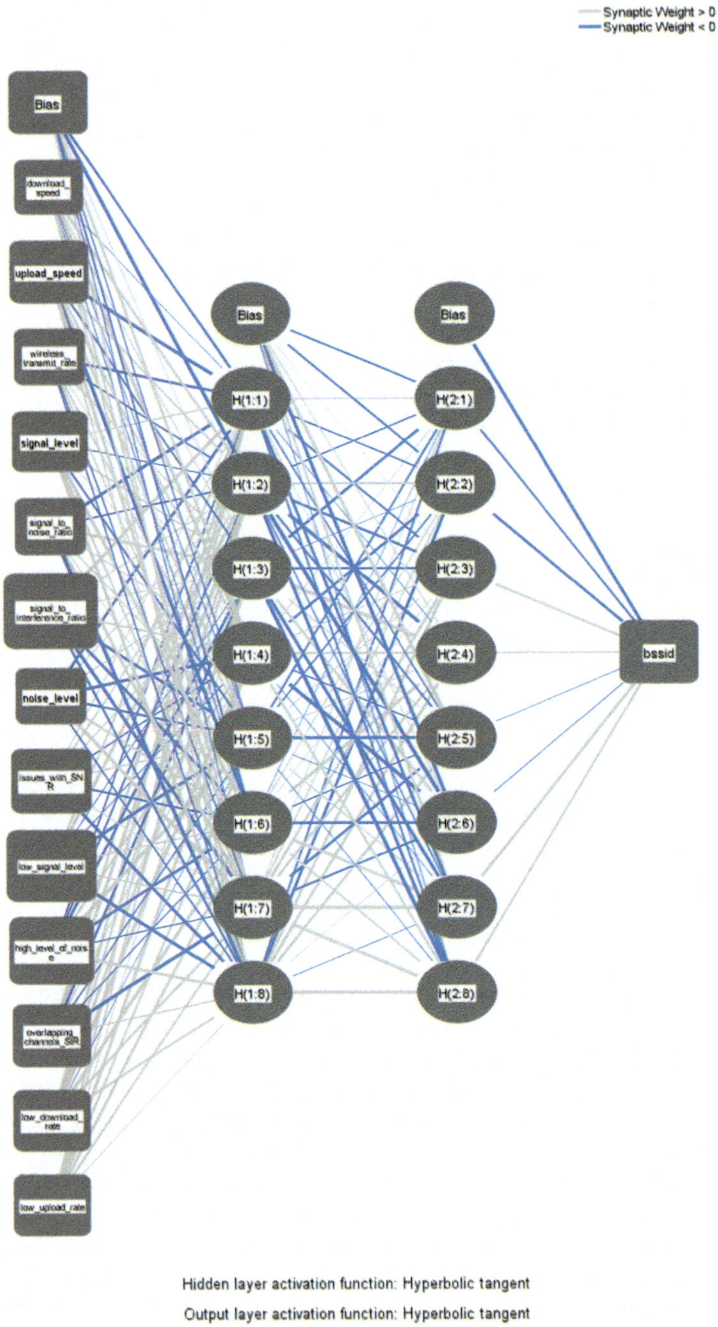

Fig. 10 Neural network based on MLP for 2 hidden layers and 8 units for the 1st and 2nd hidden layers for experimental test No. 2

Table 5 Importance level of the independent variables for experimental test No. 2

Independent variables	Importance	Normalized importance (%)
download_speed	0.028	17.6
upload_speed	0.079	49.6
wireless_transmit_rate	0.037	23.4
signal_level	0.064	40.1
signal_to_noise_ratio	0.039	24.8
signal_to_interference_ratio	0.159	100.0
noise_level	0.038	23.6
issues_with_SNR	0.088	55.1
low_signal_level	0.138	86.8
high_level_of_noise	0.110	68.9
overlapping_channels_SIR	0.077	48.2
low_download_rate	0.072	45.0
low_upload_rate	0.071	44.2

width field of the column, the alignment of the field, type of measurement and the role that the field will play in the measurement process, which is normally an output.

4.3 Performance Evaluation

In the case of experimental test No. 1, the following results are obtained, which we can see in the following Fig. 11, in which the independent variables are assessed based on the importance of normalization.

In Fig. 10, we can see that from the 0.15 importance level the parameter or independent variable issues_with_SNR obtains 100% of the normalized importance, which means that the events that occur in the Ad-Hoc wireless network have an intrinsic relationship with the relationship present between the noise and the signal in the wireless communication medium, in our case, the overlap that may occur in the waves emanating from each of the access points.

On the other hand, in the case of experimental test No. 2, the following results are obtained, which we can observe in Fig. 12 and where the covariates are assessed based on their level of importance and the normalized importance.

Indeed, in Fig. 12, we can see how the parameter signal_to_interference_ratio with an importance level of 0.15 obtains a normalization importance of 100%, which shows that the average signal interference can decrease the performance of the signals. Ad-Hoc networks. Similarly, there are two other independent variables that are key in this type of Ad-Hoc wireless network scenarios, such as low_signal_level and high_level_of_noise, with an importance level of 0.138 and 0.110, respectively and

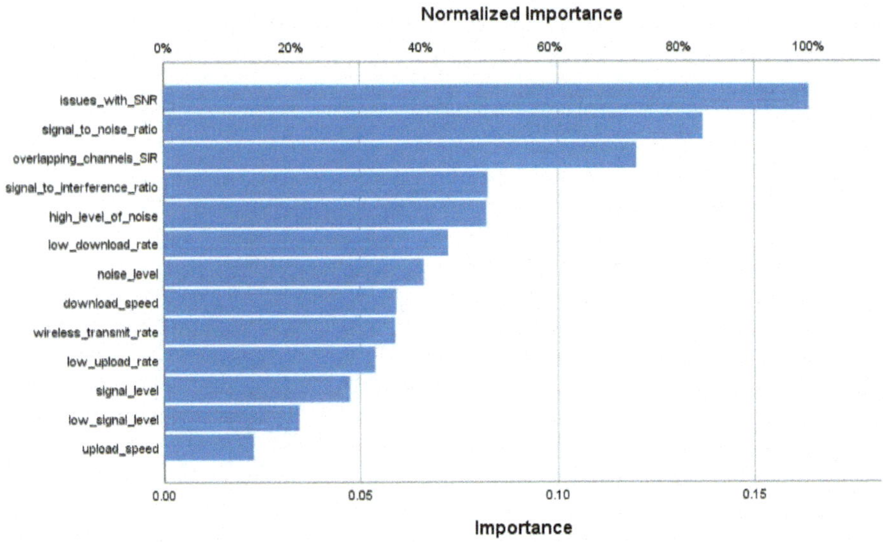

Fig. 11 Importance of normalized for experimental test No. 1

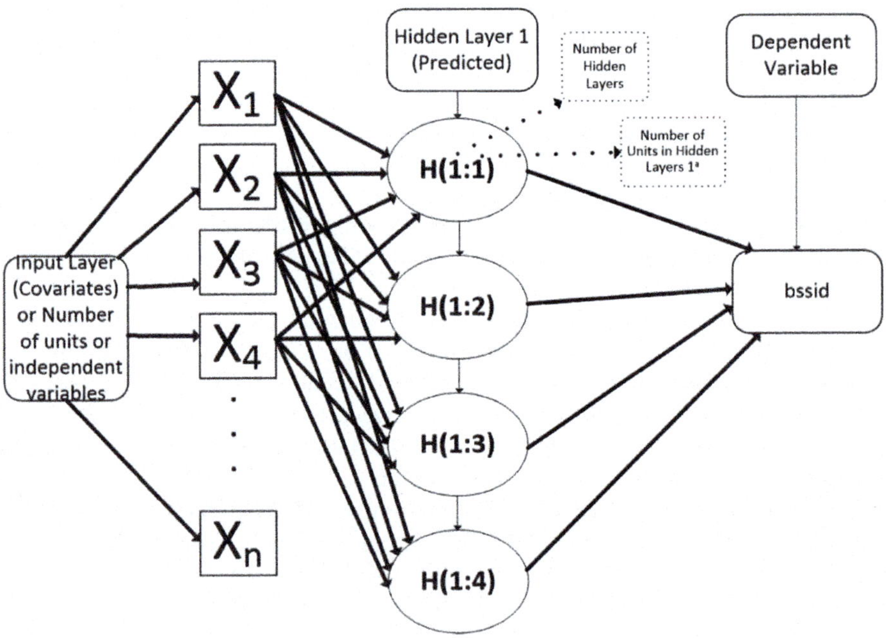

Fig. 12 Importance of normalization for experimental test No. 2

which in turn represents a normalization importance of 86.8% and 68.9%, consecutively, which shows that these parameters also affect this type of networks.

5 Discussion

In the case of experimental test No. 1, we used the following linear Eq. (3).

$$y = 0.59 + 0.81x \tag{3}$$

This linear function allows us to generate a predicted value regarding the dependent variable, bssid, which contains a single unit that corresponds to the output layer, as we can see in Fig. 13.

In Fig. 13, we can observe that in the regression line there are a total of 4 values that are above the line, which means that they are significant values that represent a regression model and that serve as a response to a set of data, which in our case, is identified by the independent variables. This allows us to quantify the relationship between one or more predictor variables and a response variable (bssid).

We also consider the residual values and their relation to the predictive values, as we can see in Fig. 14.

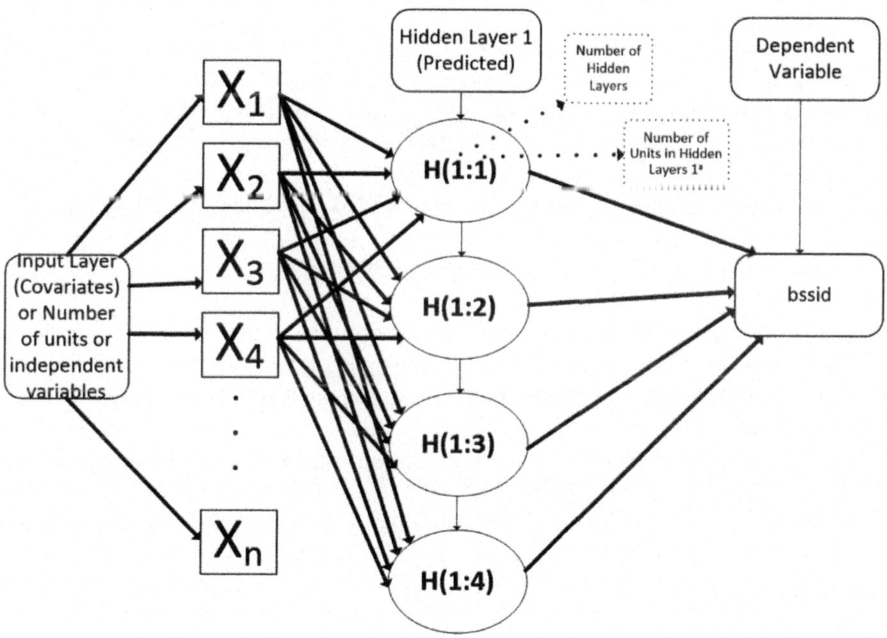

Fig. 13 Predictive value regarding the dependent variable bssid (experimental test No. 1)

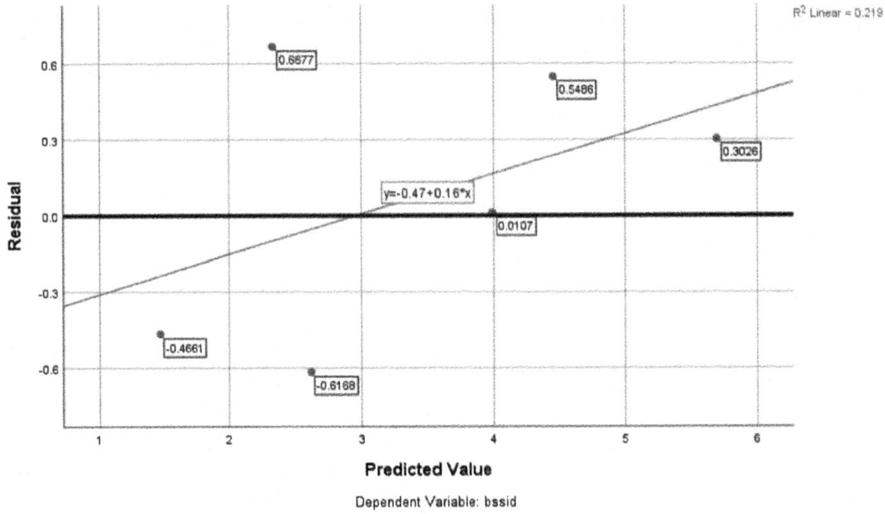

Fig. 14 Residual value with respect to the predictive values of a bssid response variable

In Fig. 14, there is the following linear Eq. (4).

$$y = 0.47 + 0.16x \tag{4}$$

At the same time, this linear function allows us to observe the relationship between the independent variables and the dependent variables. In this case, in Fig. 14, we can see that there are only 2 values that are above the regression line, which means a lower correlation between the observed values and a predicted value in the regression analysis.

In this same direction, in the experimental test No. 2, we have the following linear Eq. (5).

$$y = 2.25 + 0.35x \tag{5}$$

In Fig. 15, we can see that there are only 3 optimal values above the regression line, which tells us about the level between the independent variables and a response variable. Therefore, the regression model indicates the level of granularity that may exist with your data set.

On the other hand, we also have the relationship between residual values and predictive values as we see in Fig. 16.

In Fig. 16, we can see that the regression line is inclined in a decreasing manner, in which only 3 optimal values are observed. Thus, it seems that the relationship of one or more independent variables with respect to the dependent variable is low, or at least does not completely satisfy the response variable.

In Table 6, we make a comparison between the linear equations and the regression analysis, considering the data obtained from experimental tests 1 and 2, respectively.

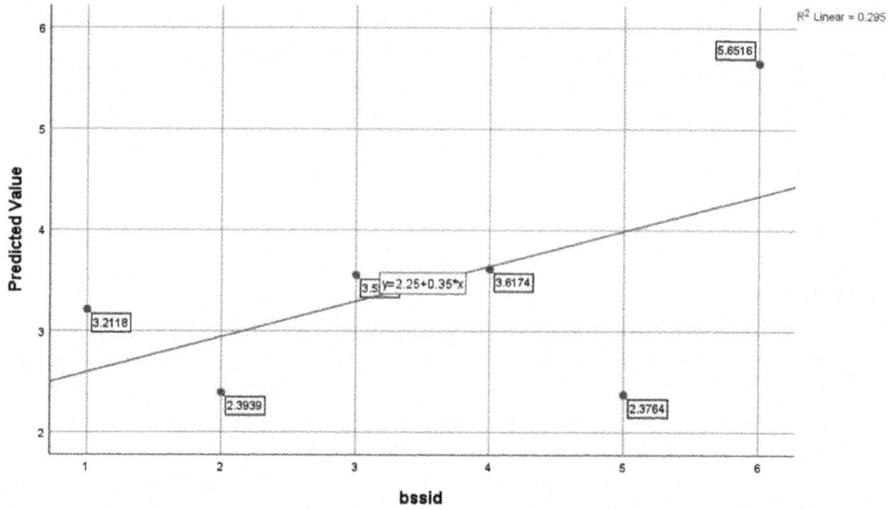

Fig. 15 Predictive value regarding the dependent variable bssid (experimental test No. 2)

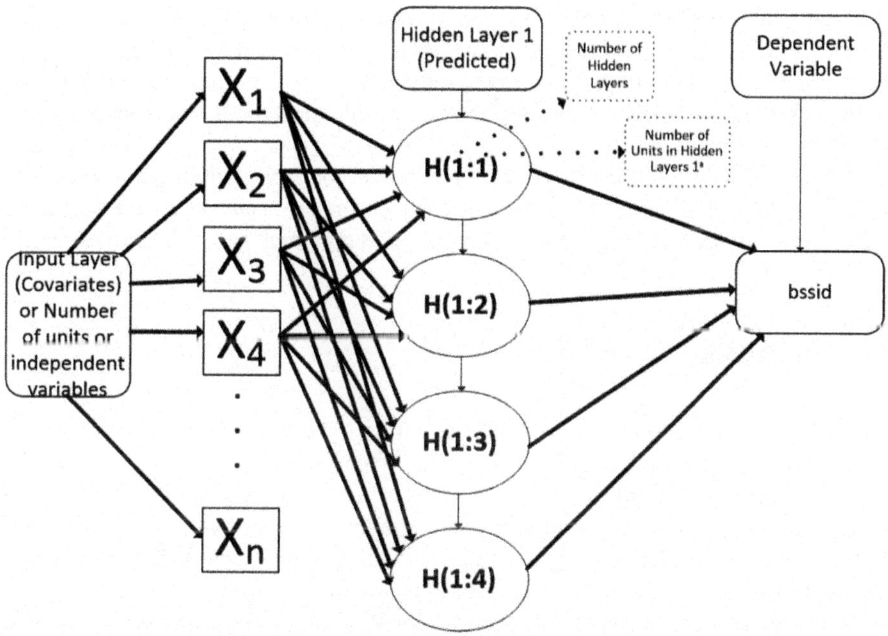

Fig. 16 Residual value with respect to the predictive values of a bssid response variable

Table 6 Linear equations versus regression analysis between experimental test No. 1 and No. 2, respectively

	Linear equation	Regression analysis (R^2 linear)	Percentage of regression analysis (%)
Experimental test No. 1			
Predicted value versus bssid	$y = 0.59 + 0.81^{*}x$	0.937	93.7
Residual versus predicted value	$y = 0.47 + 0.16^{*}x$	0.219	21.9
Experimental test No. 2			
Predicted value versus bssid	$y = 2.25 + 0.35^{*}x$	0.295	29.5
Residual versus predicted value	$y = 0.56 - 0.15^{*}x$	0.014	1.4

In Table 6, we can see that the percentage of the regression analysis is obtained by experimental test No. 1 with 93.7%, which is equivalent to 0.937 in linear regression. This is followed by the 29.5% of experimental test No. 2, with a 0.219. Likewise, experimental test No. 1 presents a better relationship of its independent variables with respect to the response variable than the results presented in experimental test No. 2.

Therefore, with 2 hidden layers and 4 elements for each with these layers, better results are obtained regarding the relationship of the covariates with the dependent variable than without using 4 hidden layers and 8 elements for each of these layers.

6 Conclusion

Generative Artificial Intelligence (GenIA) within the context of Deep Learning and through its neural networks, such as the Multilayer Percepton (MPL) and the Radial Basic Function (RBF), allows us to carry out a series of analyzes based on of a group of data, in our case known as independent variables, response variables. The contribution of this work lies in selecting those variables or parameters that are captured through a simulator suitable for scenarios that contain Ad-Hoc wireless access points in combination with Mesh networks and converting them into measurable variables that allow improving this type of scenario to starting from this set of independent variables.

As mentioned previously, 2 experimental scenarios have been used, in which tests have been carried out with MPL, using 2 hidden layers and between 4 to 8 numbers of elements for each of these hidden layers, respectively. The results show that to the extent that we increase the number of hidden layers and the number of elements for each of these hidden layers, the percentage of the Regression Analysis (linear R2) tends to decrease, which indicates that this percentage of the relationship between these independent variables and the response variable does not obtain the best optimal solution but simply a predictive or approximate value that allows providing a solution to the problem presented by the Ad-Hoc wireless network scenarios such as bandwidth, blind or dead spots and the data load balance that must be managed at a given time.

As we can see, there are a series of parameters or predictors that affect these scenarios of Ad-Hoc wireless networks, among which it is worth mentioning the average interference in the signal, the average noise in the signal, the channel overlap, the level low signal, a high noise level in the communication medium, among others.

As future work, it would be necessary to experiment with more levels of layers and number of elements per layer, for example, 4 hidden layers and 16 elements for each hidden layer based on the parameters or variables that are identified in Information Centric Networks [13]. Although for this type of experiments more computational equipment will be required to allow the results obtained to be satisfactorily managed.

References

1. Cheng, Y., Yin, B., Zhang, S.: Deep learning for wireless networking: the next frontier. IEEE Wirel. Commun. **28**(6), 176–183 (2021). https://doi.org/10.1109/MWC.001.2100005
2. Shen, W., Wang, W.: Node identification in wireless network based on convolutional neural network. In: 2018 14th International Conference on Computational Intelligence and Security (CIS), pp. 238–241. IEEE (2018). https://doi.org/10.1109/CIS2018.2018.00059
3. Guha, D.R., Patra, S.K.: Cochannel interference minimization using Wilcoxon multilayer perceptron neural network. In: 2010 International Conference on Recent Trends in Information, Telecommunication and Computing, pp. 145–149. IEEE (2010). https://doi.org/10.1109/ITC.2010.50
4. Rao, P.S., Varma, K., Satapati, R.A., Vamsidhar, E.: Multilayer perceptron based secure media access control protocol for wireless sensor networks. In: 2010 IEEE International Conference on Computational Intelligence and Computing Research, pp. 1–5. IEEE (2010). https://doi.org/10.1109/ICCIC.2010.5705769
5. Eyobu, O.S., Edwinah, K.: A deep learning-based routing approach for wireless mesh backbone networks. IEEE Access **11**, 49509–49518 (2023). https://doi.org/10.1109/ACCESS.2023.3277431
6. Canêdo, D., Romariz, A.: Intrusion detection system in ad hoc networks with artificial neural networks and algorithm k-means. IEEE Lat. Am. Trans. **17**(07), 1109–1115 (2019). https://doi.org/10.1109/TLA.2019.8931198
7. Parameshachari, B., Manjunath, T., et al.: Adaptive voting mechanism with artificial butterfly algorithm based feature selection for IDS in MANET. In: 2023 IEEE International Conference on Integrated Circuits and Communication Systems (ICICACS), pp. 1–7. IEEE (2023). https://doi.org/10.1109/ICICACS57338.2023.10099861
8. Gutiérrez, D., Toral, S.: Deep neuronal based classifiers for wireless multi-hop network mobility models. In: 2019 18th IEEE International Conference on Machine Learning and Applications (ICMLA), pp. 602–607. IEEE (2019). https://doi.org/10.1109/ICMLA.2019.00111
9. Jaton, N., Gyawali, S., Qian, Y.: Distributed neural network-based DDoS detection in vehicular communication systems. In: 2023 16th International Conference on Signal Process-

ing and Communication System (ICSPCS), pp. 1–9. IEEE (2023). https://doi.org/10.1109/ICSPCS58109.2023.10261135
10. Srinivasan, S.M., Truong-Huu, T., Gurusamy, M.: TE-based machine learning techniques for link fault localization in complex networks. In: 2018 IEEE 6th International Conference on Future Internet of Things and Cloud (FiCloud), pp. 25–32. IEEE (2018). https://doi.org/10.1109/FiCloud.2018.00012
11. Wang, Z.: The applications of deep learning on traffic identification. BlackHat USA **24**(11), 1–10 (2015)
12. Palmieri, F., Fiore, U., Castiglione, A.: A distributed approach to network anomaly detection based on independent component analysis. Concurr. Comput. Pract. Exp. **26**(5), 1113–1129 (2014). https://doi.org/10.1002/cpe.3061
13. Castillo, A.C.: An overview of integration of the virtualization of network functions in the context of information centric networks. In: 2023 3rd International Conference on Electrical, Computer, Communications and Mechatronics Engineering (ICECCME), pp. 1–7. IEEE (2023). https://doi.org/10.1109/ICECCME57830.2023.10253441

Deep Learning for Intrusion Detection on Controller Area Networks

Gokhan ALTAN, Ece HABEŞOĞLU, and Ipek ABASIKELEŞ TURGUT

Abstract The increase in the negativities experienced in traffic mobility is one of the recent problems of crowded population. With the acceleration of developments in wireless communication in recent years, it is seen that some efforts have been made to prevent traffic jams and accidents in big cities. Vehicular Ad Hoc Networks (VANET) are networks developed for purposes, such as regulating traffic flow and saving lives. HCRL-VANET, which was developed to detect and prevent security measures and threats in Inter-Vehicle Networks, monitors the behavior of vehicles and infrastructure devices and is evaluated accordingly. In this study, intrusion detection performance was evaluated using conventional machine learning and deep learning methods on the dataset obtained from data traffic circulating in HCRL-VANET networks. In the analysis, the DNN method has reached the highest intrusion detection rate of 94.31% for accuracy.

Keywords Deep learning · VANET · IDS · Fuzzy attack · DOS · DNN · ML

1 Introduction

With the advancement of computing technology, it seems that one of the biggest developments in autonomous vehicles is reducing the cost of computing opera-

Gokhan ALTAN, Ece HABEŞOĞLU and Ipek ABASIKELEŞ TURGUT are contributed equally to this work.

G. ALTAN (✉) · E. HABEŞOĞLU · I. ABASIKELEŞ TURGUT
Department of Computer Engineering, Iskenderun Technical University, Hatay, Türkiye
e-mail: gokhan.altan@iste.edu.tr

E. HABEŞOĞLU
e-mail: ece.habesoglu.lee23@iste.edu.tr

I. ABASIKELEŞ TURGUT
e-mail: ipek.abasikeles@iste.edu.tr

G. ALTAN
R&D, Deeptech Engineering Trade Co. Ltd., İskenderun, Hatay, Türkiye

G. ALTAN and I. ABASIKELEŞ TURGUT (eds.), *Deep Learning in Ad-Hoc Wireless Networks*, Studies in Big Data 172,
https://doi.org/10.1007/978-3-031-86075-1_5

tions [1]. The importance of resource needs increases with the increasing global population. Since communication is important in terms of saving time, energy and money, VANET (Vehicular Network) was developed by researchers [2]. VANET also aims to increase communication efficiency in order to provide communication, that is, to provide return and exchange of information [3]. A description of SDN (Software Defined Networking)—VANET applications in different wireless communication fields such as Internet of Things (IoT) and VANET is presented and the focus is on examining and comparing SDN-VANET studies on many parameters [4].

Deep learning (DL) is a sub-field of machine learning used to learn complex data representations using mathematical models called artificial neural networks. DL is usually performed using multilayer artificial neural networks (Deep Neural Networks). DL has become a necessity today, especially for Internet of Things technology and root applications of communication. In [5], security vulnerabilities were created for multiple IoT devices in order to enable sensor systems to access the internet, remote control capabilities, communication and coordination between systems. The aim was to design a DL architecture for the implementation of secure IoT applications and to evaluate the performance and robustness of Deep-Learning algorithms applied for IoT-based invasion detection [5].

DL uses many layers of non-linear processing units for feature extraction and feature learning. Each successive layer takes the output from the previous layers with a fully connection on inputs [6]. Architectures can be trained either supervised (such as classification) or unsupervised (such as pattern analysis). DL covers methods that enable high-level abstractions from data. Deep neural networks (DNN), which are artificial neural network architectures for DL, are actually artificial neural networks with multiple hidden layers. 3-layer feedforward networks (with a single hidden layer) used in practice have a structure that can be described as shallow. The reason why it is called shallow is that the features of the data are calculated with only one single hidden layer. It is possible to calculate more complicated features with deep networks. Therefore, each hidden layer has a nonlinear transformation of the summation of previous layer. In this way, deep networks can learn more complicated functions, which provides stronger representation ability [7]. DL is used in automatic driving technologies such as autonomous vehicles and driver assistance systems. By the capabilities of computer vision, it is used to detect objects, recognize traffic signs, follow other vehicles, drive safely, and more.

Vehicular routing in VANETs is used to manage the movements of vehicles on a network and determine the most efficient routes. Vehicle routing with deep reinforcement learning is an approach that enables vehicles to make dynamic and real-time routing decisions. This approach provides great benefits, especially in intelligent transportation systems (ITS) and vehicle-to-vehicle (V2V) communication.

RSU (Road Side Unit) are hardware units placed on the side of roads as part of vehicle-to-vehicle communication (V2X) systems. They are used to communicate with surrounding vehicles and infrastructure. In VANET, there are also vehicle-to-roadside unit (RSU) applications in addition to vehicle-to-vehicle applications. Studies using RSU have gradually increased and become applicable on ML technologies [8].

Problem Statement: Ensuring efficient data transmission between vehicles is a significant challenge that requires a reliable routing protocol. In this paper, DL and ML algorithms are utilized to ensure VANET achievements, improve routing protocols, increase security and maximize driving efficiency by intrusion detection. Solutions have been proposed to increase efficiency by using classification techniques, including DNN and ML techniques.

With the increase in information and delivery of packet rates on communication in the automobile industry, security vulnerabilities are emerging in VANETs. These networks are used to enable data communication between electronic components, but often have security vulnerabilities. In particular, protocols such as Controller Area Network (CAN) do not include information about source and destination ID when sending broadcast messages. It allows malicious attackers to intervene in the system and affect the vehicle's functions. Systems such as Electronic Control Units (ECUs) are adapted to control functions within the vehicle, but ECUs are also vulnerable to attacks. Remote code execution attacks can be carried out via Bluetooth interfaces, which can affect the vehicle's behavior. The vulnerabilities can lead to various types of attacks. The attacks can be carried out by exploiting weaknesses in VANETs. In this context, new methods need to be developed to detect and protect vulnerabilities the communications in VANET. In this study, six conventional machine learning and DNN were trained to detect intrusion attacks such as Denial of Service (DOS) and fuzzy attacks.

2 Related Works

VANETs are vulnerable to security threats. IDS (Intrusion Detection System) has been chosen to detect anomalies and malicious activities in the networks. Tejasvi et al. needed better anomaly detection frameworks that can handle this unexpected situation since there may be many types of anomalies in the network. They proposed a VANET anomaly detection framework focusing on DNNs using sequence reconstruction and thresholding algorithm [9].

Clustering in VANETs allows tools to be organized into specific groups. This optimizes data transmission and network management processes. Khalid Kandali et al. proposed a cluster-based protocol for VANET by combining MSSP and CHN with the modified K-means algorithm. Initial cluster heads were selected using MSSP and CHN, and after determining the number of clusters, more stable and reliable cluster heads were obtained by avoiding random selection. Using the modified K-means algorithm, vehicles were assigned to clusters based on the link reliability model. This model takes into account the mobility of vehicles and the dynamic structure of the network. In the maintenance step, an appropriate weight function was used and new cluster heads were selected, taking into account parameters such as the node's buffer size, speed and node degree. This approach makes data transmission in VANETs more efficient and reliable, while reducing traffic congestion and aiming to guarantee the stability of clusters even at high density and mobility [10].

Traffic management in VANETs is important to optimize traffic flow and reduce traffic congestion. Traffic light timing optimization must be adaptable, taking into account the number of vehicles on the road and the options available to direct vehicles to their destinations. Ritesh et al. developed the first model based on DL to solve the traffic congestion problem in the urban environment. The target was trained using a network-based deep double-Q learning mechanism by providing adaptive traffic light control, the average waiting time was reduced by up to 91.7% [10].

GIDS, a GAN-based IDS, was used for the in-vehicle network. GIDS affects the performance and speed factor. In the GAN-based study, CAN network traffic was identified, and it was a difficult task to distinguish abnormal traffic due to intentional attacks. The proposed model has been shown to help increase accuracy in detection [11]. The security features of the CAN protocol are intended to make vehicles vulnerable to different attacks. It was aimed to make the types of computers vulnerable and used different types of algorithms and classes [12].

In recent days, autonomous cars pose a wide range of risks. The aim of their study is to design an intelligent anomaly detection system and propose an IDS that can detect anomalies [13]. Kang and Kang [14] trained in-vehicle packet management using DBN classification. The system takes into account all possibilities and categorizes normal and attack. They reached a classification success of 97.8% through the VANET IDS dataset. Song et al. experimented with multiple classifiers, achieved high classification performances with the help of deep neural networks to distinguish attacks in in-vehicle network traffic [15].

3 Materials and Methods

3.1 Software-Defined Networking (SDN)

SDN, Software-Defined Networking, is a technology that moves network management and controls from traditional hardware-based methods to a software-based approach. SDN emphasizes the separation between the control plane and the data plane of the network. Thus, performing network management from a central control point. The theoretical foundations of SDN cover a broad field of research covering topics, such as communication protocols, algorithms, and security.

SDN is a logically centralized network framework for the control plane. SDN architecture is based on the principle of separating the control plane from the data plane. SDN communication protocol is standardized by ONF (Open Network Infrastructure Foundation) and provides a two-way communication tunnel for rule insertion and packet exchange between controller and switch. On the other hand, the data plane represents the packet forwarding medium between network switches [16].

3.2 VANET Communication Types

VANET is defined as a type of network that supports wireless communication between vehicles. These networks are a temporary communication infrastructure where vehicles can communicate with each other and with the road infrastructures, such as traffic signs, traffic lights, roadside units, etc.). VANETs have significant potential for a variety of applications such as traffic safety, traffic management and quality of service by supporting direct and indirect communication between vehicles and between vehicles and infrastructure. Traditional VANET services include vehicle and road safety management and traffic safety, and the effectiveness of traffic services is designed to improve traffic flow and benefit the region [4]. V2P (Vehicle to Pedestrian), V2V (Vehicle to Vehicle) and V2I (Vehicle to Infrastructure) communications are three basic subcategories of VANETs. They represent different communication scenarios between vehicles, vehicles and roadside infrastructure, and vehicles and infrastructures. V2P communication allows vehicles to communicate with pedestrians, while V2V communication refers to direct communication between vehicles. V2I communication allows vehicles to communicate with roadside infrastructure, usually through roadside hardware units (such as RSUs). RSU's are also called routers because they work like a router in V2V communication on the path and interact with other network devices [4]. These communication types support the core functionality and purposes of VANETs and are often operated in various applications such as traffic flow management, security and quality of service [3].

3.3 HCRL-VANET

HCRL (Hacking and Countermeasure Research Lab) carries out research and development activities in the field of cybersecurity. HCRL continues its work to develop protection techniques against cyber-attacks, evaluate cyber threats and improve cybersecurity technologies [17]. The connection of HCRL with VANET may address issues such as the security of vehicle-to-vehicle communication, protection measures against cyber-attacks, and the reliability of the VANET infrastructure. Many recent papers have aimed to develop solutions to cope with the cybersecurity challenges of VANET and increase security on the vehicles' communication network.

It generally includes security research and can use various methods to identify security threats, vulnerabilities and risks faced in VANET. The output could provide cybersecurity experts, researchers, and industry accompanies with usable data to improve the security in VANET.

3.4 Intrusion Detection on HCRL-VANET

HCRL-VANET intrusion detection is an advanced security technology developed to ensure security in modern vehicle networks [18]. HCRL detected VANET anomalies and attacks using Deep Reinforcement Learning and machine learning methods. VANETs enable vehicles and RSUs to communicate with each other dynamically. These networks are particularly vulnerable to security threats. Therefore, it requires an effective IDS.

HCRL-VANET is an algorithm for intrusion detection in VANETs and aims to identify certain types of instructions while monitoring and examining communication traffic in the network. Instructions are often represented by certain patterns or certain features in network traffic. HCRL-VANET can use machine learning or artificial intelligence techniques to detect these instructions using data traffic and packet types.

The main goal of HCRL-VANET is to detect intrusions in network traffic and determine whether these instructions are either normal or abnormal. This is important to detect and prevent security breaches in VANET networks. An intrusion detection system can identify and take precautions against malicious behavior or potential attacks on the network.

Wireless communication is provided between VANETs, vehicles and infrastructures. Each communication technology is different from each other due to unique features and high mobility. However, it remained vulnerable to various security attacks due to the lack of a central infrastructure. It poses a serious threat to the safety of road traffic [10].

3.5 Dataset

Abdulaziz et al. proposed a security breach detection system that detects attacks on vehicles. They exploited two algorithms based on KNN and SVM to detect DoS and fuzzy attacks. During the analysis process, they utilized two vehicle hacking datasets provided by HCRL: "DoS data set" and "fuzzy data set". Each attack lasted 3–5 s. 30–40 min of CAN traffic is required for each data set. The DoS dataset contains 3,665,771 messages (3,078,250 normal messages and 587,521 injection messages). It has twelve features. Fuzzy dataset contains 3,838,860 rows of messages (2,759,492 normal messages and 1,079,368 injection messages) with twelve features, as well. In DoS attacks, the injection message contains the CAN ID '0×000' while in

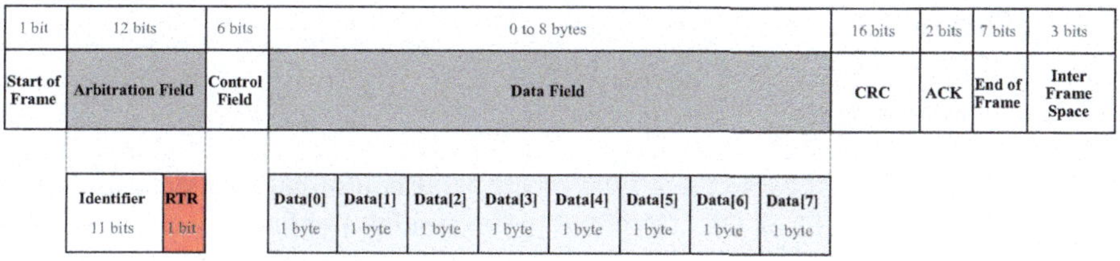

Fig. 1 CAN data frame structure [12]

Table 1 Data packet information in the VANET traffic [12]

TIME STAMP	CAN ID	DLC	DATA [0]
DATA [1]	DATA [2]	DATA [3]	DATA [4]
DATA [5]	DATA [6]	DATA [7]	Flag

fuzzy attacks, the injection messages are pseudo-random CAN ID and DATA values. Figure 1 shows the description of information about flowing data. The structure of the two datasets is similar, but they represent different types of attacks. The DoS dataset represents DoS attacks that inject a CAN ID with "0000" message on every 0.3 milliseconds, where "0000" is the most dominant. However, the fuzzy attack dataset injects completely random messages into the CAN ID and data values on every 0.01 ms. The eight feature on flowing data are shown in Table 1.

In this study, we disposed attack dataset and fuzzy dataset for the analysis to identify intrusions in VANET. These data sets were obtained by connecting CAN traffic from real vehicles via the OBD-II port. The data was preprocessed and then utilized with DNN, MLP, Random Forest, KNN and M5P algorithms in the classification stage. Attack dataset contains a total of 249,108 rows and 11 features. The fuzzy data set contains 177,818 rows and 11 features. In this study, a data set with a total of 426,925 rows was obtained by combining the attack and fuzzy data sets. The features are shown in Table 2. The "class" column stands for the labels attack and fuzzy for binary classification.

Table 2 Experimented features

ID	DLC [1]	DLC [2]	DLC [3]
DLC [4]	DLC [5]	DLC [6]	DLC [7]
DLC [8]	DLC [9]	Class	–

3.6 *Classifiers*

3.6.1 Deep Learning

DNN (Deep Neural Network) is a type of artificial neural networks and is widely used model in the field of DL. DNNs is the extended architecture of an artificial neural network architecture consisting of many hidden layers and big number of neurons. It is often selected to model complicated relationships over large amounts of data. DNNs have three or more layers: input layer, one or more hidden layers, and an output layer. Each layer is connected to each other with different structures, such as fully-connection with adaptive optimization and factorization techniques.

DNNs are often trained with large amounts of data. Thus, it enables increasing the generalization capability. During training, a backpropagation algorithm is used to reduce the loss and error between the actual output and the predicted output of model.

3.6.2 Machine Learning

MLP (Multilayer Perceptron) is one of the most basic and widely used types of artificial neural networks. MLP has a structure that includes at least one input layer, at least one hidden layer and one output layer. Each layer consists of fully connected nodes. Each node calculated the actual value of nodes using randomly defined weights, input data, and activation function as an output. Training of MLP is carried out using the backpropagation algorithm on the data set [19]. This algorithm compares the model's predictions to the actual labels, then updates the weights in the model to minimize the output error. The most commonly selected backpropagation algorithm is gradient descent algorithm [20].

Decision tree is one of the supervised learning methods that is used to solve classification and regression problems. Decision trees have a tree-like structure containing many decision nodes and outcome nodes. Each decision node performs a test on a feature and divides the data into subsets based on this test result. The branching of the tree continues by splitting the data set according to successive combinations of these tests. This splitting process aims to maximize the information gain in the dataset with the help of entropy [20].

Random forest is a machine learning algorithm used to solve classification and regression problems. Random forest is a combination of many decision trees and creates a forest consisting of these decision trees in the training. It is an algorithm that has a low tendency to overfitting, performs well and is effective on high-dimensional data sets. It is also often used to determine and understand the responsibility and importance of features [20].

Random Tree is one of the basic building blocks of the random forest algorithm. It consists of a single decision tree, is created using techniques such as random sampling and random feature selection in random forest. The main idea of the random tree is to select random features at each node and find the best splitting point based on those features. It increases the diversity of the tree and reduces overfitting. Consequently, the predictions of random tree are combined and an averaged result is produced [20]. M5P is a machine learning algorithm used to solve regression problems. The M5P algorithm creates a series of decision trees on the data set. However, unlike a standard decision tree, M5P trees can handle both continuous and categorical data types, simultaneously, which makes M5P usable in regression problems [20].

KNN (k-Nearest Neighbors) is a machine learning algorithm, which is used to solve classification and regression problems. The basic idea is to use the labels or values of its k number of closest neighbors to predict the label or value of a data point. The KNN algorithm has a simple and understandable structure, which is easy to implement. Additionally, there is no model training phase as there is no training process [19].

4 Experimental Results

In modern systems, such as connected vehicles, intelligent intrusion detection systems have become an important security issue for the researchers. The vehicles have become the target of various attacks that affect the performance of network, threaten public and private property, and harm road safety [17].

In this study, we use two datasets of Attack and Fuzzy attack types. We preprocessed using data manipulation. Then, we applied KNN, MLP, M5P, DNN, decision tree, and random forest algorithms. The majority of the classifiers yielded good identification performance results, whereas the DNN outperformed all algorithms. Table 3 presents a comparative analysis on the achievements of conventional machine learning applications.

Table 3 The achievements for traditional machine learning algorithms

Classifier	Evaluation metrics			
	mae	rmse	rae	rrse
KNN (k = 1)	0.0519	0.1645	10.6811	33.3712
KNN (k = 2)	0.0524	0.1647	10.7798	33.419
KNN (k = 3)	0.0529	0.1655	10.8914	33.5645
Random Forest	0.0521	0.1624	10.7149	32.9507
M5P	0.0603	0.1706	12.4132	34.6107
M5P	0.0598	0.2444	12.3093	49.5766
Random tree	0.0517	0.1640	10.6339	33.2627
Random tree	0.0517	0.1637	10.6315	33.2099
MLP	0.3149	0.3926	64.7926	79.6329
MLP	0.2920	0.3937	60.0757	79.8615

Whereas Karthiga et al. performed an intrusion detection system using KNN and SVM algorithms with effective achievements in detecting DoS and Fuzzy attacks in VANETs, we reached the superior performance using DNN after data manupulation. It was demonstrated by using a wider range of algorithms on hyper-parameter selection for iterative models. This comparison makes an important contribution to understand how different algorithms and models can perform an important role in detecting security breaches in vehicles. We selected the mean square error (mse), mean absolute error (mae), root mse (rmse), root absolute error (rae), root relative squared error (rrse), and accuracy as the assessment of the performances on the experimented algorithms and DL architectures.

As a result of model training with the dataset, 4 rectified linear unit (relu) activation functions and sigmoid activation function were used at the output. In training, the mean square error (mse) loss function was used with two layers with 64 and 32 hidden units, 50 epochs and a batch size of 3. The loss value of the model in the test data was 0.4169 and the accuracy rate was 58.72%, whereas the loss value in the training data was 0.4169 and the accuracy rate was 58.31%.

When cross validation was entered as 5 folds, the correlation coefficient value was determined as 0.9442 using data analysis with random forest. Intrusion detection error values of 0.052, 0.1624, 0.1071 and 0.3295 were reached for mae, rmse, rae and rrse in 426924 samples. The dataset is split as percentages of 90%–10% for training-testing, respectively. On the hand, the available data was split so that 90% and 10% for training and testing, the correlation coefficient value was determined as 0.9442. Similarly, intrusion detection error values of 0.052, 0.1621, 10.72% and 32.92% were reached for mae, rmse, rae and rrse in 42692 samples, respectively.

The cross validation was set to 5 folds, the correlation coefficient value was determined as 0.9431 for data analysis with random tree algorithm. Intrusion detection error values of 0.051, 0.164, 0.1063 and 0.3326 were reached for mae, rmse, rae and rrse in 426924 samples, respectively. Additionally, the dataset is split as percentages of 80%–20% for training-testing. The correlation coefficient value was determined as 0.9436. Intrusion detection error values of 0.0514, 0.1632, 0.1057 and 0.3311 were achieved for mae, rmse, rae and rrse in 85385 samples, respectively. Otherwise, for a splitting percentage of 90%–10% for training-testing, the correlation coefficient value was determined as 0.9433. Intrusion detection error values of 0.0518, 0.1634, 0.1066 and 0.3319 were reached for mae, rmse, rae and rrse in 42692 samples, respectively.

We experimented for the different values of k for data analysis with KNN. For a splitting percentage of 90%–10% for training-testing, the correlation coefficient value was determined as 0.9429. Intrusion detection error values of 0.0519, 0.1639, 10.69% and 33.29% were reached for mae, rmse, rae and rrse in 42692 samples, respectively. On the other hand, the dataset is split using the percentages of 80%–20% for training-testing, respectively; the correlation coefficient value was determined as 0.9436. The error values of 0.0514, 0.1631, 0.1058 and 0.3310 were reached for mae, rmse, rae and rrse in 85385 samples, respectively. The splitting value is set to the percentages of 70%–30% for training-testing, respectively; the correlation coefficient value was determined as 0.9425 and intrusion detection errors of 0.052,

0.1647, 0.1070 and 0.3342 were reached for mae, rmse, rae and rrse in 128077 samples, respectively. When the available data was separated with a folding value of 4 on the cross validation, the correlation coefficient value was determined as 0.9425, intrusion detection errors of 0.052, 0.1648, 0.1068 and 0.3342 were achieved for mae, rmse, rae and rrse in 426924 samples, respectively.

In the data analysis with M5P algorithm, the folding value was set to 5 on the cross validation. The correlation coefficient value of 0.8772 was reached using M5P. We reached the intrusion detection errors of 0.0598, 0.2444, 0.1230 and 0.4957 for mae, rmse, rae and rrse in 426924 samples, respectively. We additionally experimented with splitting the dataset into training and testing folds with the percentages of 70%–30% for training-testing, respectively. We achieved the correlation coefficient value of 0.9382 and identification error values of 0.0603, 0.1706, 0.1241 and 0.3461 for mae, rmse, rae and rrse in 128077 samples, respectively.

Dense is one of the basic neuron used in NN. It represents for the layer and nodes in many classification algorithm with a most common types. Dense layers are also known as fully connected layers in DNN, because of each neuron is connected to all neurons in the adjacent layer. We set dense layer size to 2–3 for the hidden layers in the analysis.

Activation functions are mathematical functions that enable neurons to process input data and produce output in DL and artificial neural networks (ANN). We utilized the popular activation functions, including relu, sigmoid, tanh, and softmax for training and output. Output layer uses 1 neuron and sigmoid activation function (for binary classification). We applied batch normalization that is placed after each dense layer and normalizes the data to speed up the training process with a batch size of 32. Dropout factorization rate of 0.2 that randomly disables number of 20% at certain neurons to prevent overfitting. In the training of classifiers, we selected adam optimizer with a learning rate of 0.001 for 50 epochs. The output layer was fed with binary cross entropy loss function. Table 4 presents a comparative analysis on the achievements of conventional machine learning applications.

A confusion matrix is a table used to evaluate the performance of a classification model. Summarizes the relationship between the model's predictions and actual values to evaluate the system performance, statistically [19]. It consists of four basic components:

- True Positive (TP): "Attack" samples that the model correctly classifies as "attack".
- True Negative (TN): "Fuzzy attack" samples that the model correctly classifies as "fuzzy attack".
- False Positive (FP): "Fuzzy attack" samples that the model incorrectly classifies as "attack".
- False Negative (FN): "Attack" samples that the model incorrectly classifies as "uzzy attack".

We calculate accuracy, precision, and recall statistical evaluation metrics for assessment of the proposed architectures. Accuracy stands for the ratio of the model's

Table 4 The achievements for traditional deep neural networks models

DNN	Epoch	Evaluation metrics (%)		
		Accuracy	Precision	Recall
256-256	100	94.29	99.72	86.76
128-128-128	50	94.26	99.55	86.58
64-96	64	94.06	99.18	86.44
128-256	16	94.15	99.04	86.78
64-64-64	50	94.15	99.62	98.86
64-64-64	30	94.06	99.61	86.06
64-128-256	50	94.27	99.47	86.69
32-64	50	93.72	98.87	85.88
32-16-32	100	92.77	99.67	98.68
16-64-16	40	92.05	99.91	80.95
128-256-16	200	94.27	99.68	86.49
32-16-64	20	93.45	98.61	85.45
256-256-256	100	94.31	99.40	86.86
32-32-32	100	94.09	99.49	86.32
16-16-256	100	93.46	99.07	85.02
16-256-64	30	93.31	98.41	85.29
32-16-16	64	93.83	99.12	85.91

correct predictions to the total predictions. Recall is the ratio of true positive predictions to total positive. On the other hand, precision, true positive predictions, which is expressed as a ratio of true positive predictions and false negative predictions to the total number of test samples [21].

Confusion matrix and FPR (False Positive Rate) for the receiver operating characteristic curves are presented in Figs. 2, 3, and 4.

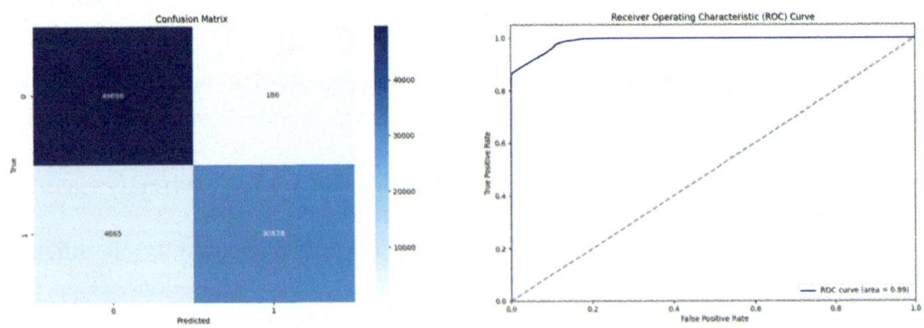

(a) (b)

Fig. 2 Confusion matrix (**a**) and ROC (**b**) for DNN model with 256-256-256 neurons

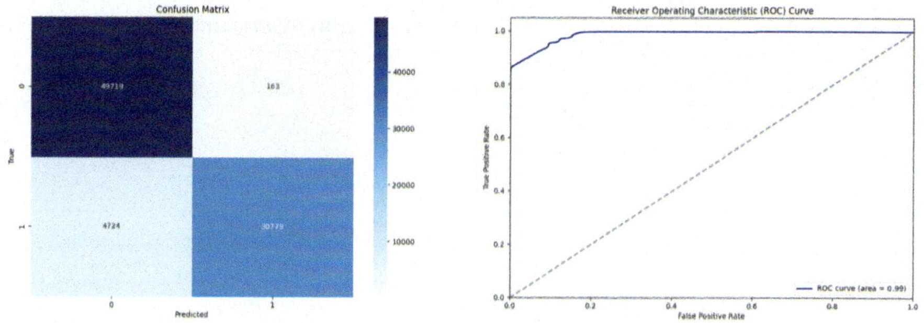

(a) (b)

Fig. 3 Confusion matrix (**a**) and ROC (**b**) for DNN model with 64-128-256 neurons

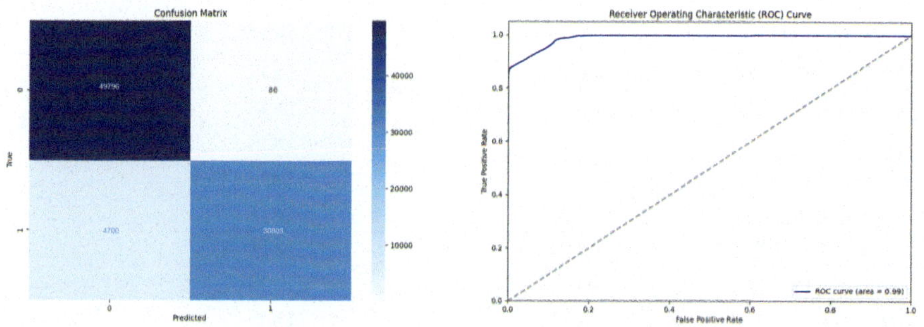

(a) (b)

Fig. 4 Confusion matrix (**a**) and ROC (**b**) for DNN model with 256-256 neurons

5 Discussion and Conclusion

The number of connected vehicles in Vehicle Ad-Hoc Networks is increasing and is a step towards realizing the goal of Intelligent Transportation System (ITS). In this study, a VANET anomaly detection framework based on DNNs is proposed using sequence reconstruction and thresholding algorithm. In this framework, DNN architectures are deployed at roadside units (RSUs) to receive broadcast vehicle data and execute anomaly detection tasks to classify a particular set of messages as abnormal or valid. In this proposed IDS architecture, various DNN architectures were implemented and their performances were compared using statistical evaluation metrics. Performance of the proposed architectures has also been compared with state-of-the-art in related IDS databased. To the best of our knowledge the best-IDS detection performance has been achieved DL-based schemes with an accuracy of 98%, providing a huge improvement over the established benchmark on VANETs [9].

Smart city growth requires information and communications technology to increase urban sustainability, but faces critical traffic congestion and vehicle classification issues among traffic. It is crucial to dynamically change traffic lights on

Table 5 Comparative evaluation of the state-of-the-art in terms of attack identification accuracy and attack type

Related works	Accuracy (%)	Attack type
Seo et al. [11]	99.50	Fuzzy
Kang et al. [14]	97.80	Fuzzy
Song et al. [15]	99.95	Fuzzy
Alsaade et al. [22]	99.00	Fuzzy
Martinelli et al. [23]	85.00	Fuzzy
Song et al. [15]	100	DOS
Hind Bangui et al. [24]	97.90	DOS
Martinelli et al. [23]	97.40	DOS
Naqvi et al. [12]	98.10	DOS
Mahmood et al. [25]	97.10	DOS
Seo et al. [11]	96.80	DOS
Karthiga et al. [26]	98.50	DOS
Alsaade et al. [22]	100	DOS
Faisal et al. [13]	96.90	DOS
Mahmood et al. [27]	96.70	DOS
Nayyar et al. [28]	97.10	DOS
Shams et al. [29]	96.50	DOS
Almi'ani et al. [30]	96.90	DOS
Nie et al. [31]	96.30	DOS

the road network to reduce vehicle delay and avoid traffic congestion in smart cities. Traffic light modification must be adaptable, taking into account the number of vehicles on the road and the options available to direct vehicles to their destinations. Table 5 presents the complete comparison on the state-of-the-art. Kumar et al. proposed a recent DL-based model to solve the traffic congestion problem in urban environment [32]. The model classifies the type of vehicles on the road and assigns different vehicle weights. The role based weight determination is the concluded side of their proposal. If there is no vehicle, a weight value of 0.0 is assigned, 1.0 for light vehicles, 2.0 for medium-weight vehicles and 3.0 for heavy vehicles. The proposed work is trained using deep double-Q learning mechanism based on experience repetition and target network on DSRC protocol for wireless communication. They utilized the simulation metrics using SUMO (Simulation in Urban Mobility) software with a randomly generated data. The results show that the traffic light of a particular traffic intersection becomes adaptable in accordance with the above-mentioned objectives. Their proposal efficiently reduced the average waiting time at road intersections by up to 91.7% [32]. Seo et al. reached 99.5% and 96.8% success rates as a result of fuzzy and DOS attacks, respectively [11]. Kang and Kang achieved an attack detection rate of 97.80%, false positive error value of 1.6%, and false negative rate of 2.8% results on the dataset [14]. However, the authors ignored features of in-vehicle network traffic, such as sequential behavioral. Song et al. presented attack identification accuracy rates of 99.95% and 100% for fuzzy and DOS attacks, respectively [14].

IDS system covers real-world traffic containing attack and non-attack data from various vehicle nodes in the VANET system. Attacks in this real-time dataset consist of Brute Force attacks, Heartbleed attacks, botnet attacks, DoS attacks, DDoS attacks, web attacks and penetration attacks. In [26], IDS-VANET system achieves a sensitivity rate of 98.27%, a specificity rate of 98.92%, and an accuracy rate of 98.60% with 1.52 s detection time. The average accuracy of the proposed IDS model is about 98.6%. Their proposal, KIDS-UIDS method, which is based on the methods in [12, 13, 24, 25, 27–31], were compared to detect various attacks in the VANET environment. The conventional methods on IDS attack detection presented the effectiveness of the smart IDS system can be verified on VANET technologies [26]. Martinelli et al. proposed a discussion with a precision ranging from 0.85 to 1 utilized the fuzzy NN algorithm in the identification of attacks targeting CAN protocol [26]. It reached attack identification accuracy rates of 85% and 97.4% for fuzzy and DOS attacks, respectively. Alsaade et al. achieved an attack identification accuracy rate of 99.98% using the car-hacking dataset and an attack identification accuracy rate of 98.09% using the UNSW-NB15 dataset. The proposal reached the highest accuracy rates of 99% and 100% for fuzzy and DOS attacks, respectively [22].

DNN allows creating various representations of input data using different layer models. The number of layers and neurons in each layer allows learning features and passing them to the next hidden layer using unsupervised methods. Using time-frequency features as input to the DNN thus defines the relational parameters and determines the associated weights in pre-training. The proposed DNN models separated fuzzy and DOS attacks with rates of 94.31%, 99.40%, and 86.86% for accuracy, precision, and recall, respectively, Since DNN has an iterative architecture with random weight initialization, it is possible to reach higher results for various hyper-parameter tuning and adaptive optimization techniques.

Acknowledgements The authors thank Deeptech Engineering Company (www.deeptechmuhendislik.com) for providing full access to GPU servers for machine learning and deep learning analysis.

References

1. Lee, M., Atkison, T.: Vanet applications: past, present, and future. Veh. Commun. **28** (2021), https://doi.org/10.1016/j.vehcom.2020.100310
2. Mahi, M.J.N., Chaki, S., Ahmed, S., Biswas, M., Kaiser, M.S., Islam, M.S., Sookhak, M., Barros, A., Whaiduzzaman, M.: A review on VANET research: perspective of recent emerging technologies. IEEE Access **10**, 65760–65783 (2022). https://doi.org/10.1109/ACCESS.2022.3183605
3. Karabulut, M.A., Shah, A.F.M.S., Ilhan, H., Pathan, A.S.K., Atiquzzaman, M.: Inspecting VANET with various critical aspects—a systematic review. Ad Hoc Netw. **150** (2023). https://doi.org/10.1016/j.adhoc.2023.103281

4. Al-Heety, O.S., Zakaria, Z., Ismail, M., Shakir, M.M., Alani, S., Alsariera, H.: A comprehensive survey: benefits, services, recent works, challenges, security, and use cases for SDN-VANET. IEEE Access **8**, 91028–91047 (2020). https://doi.org/10.1109/ACCESS.2020.2992580
5. Altan, G.: SecureDeepnet-IoT: a deep learning application for invasion detection in industrial internet of things sensing systems. Trans. Emerg. Telecommun. Technol. **32** (2021). https://doi.org/10.1002/ett.4228
6. Ozcanli, A.K., Yaprakdal, F., Baysal, M.: Deep learning methods and applications for electrical power systems: a comprehensive review. Int. J. Energy Res. **44**, 7136–7157 (2020). https://doi.org/10.1002/er.5331
7. Özdemir, D.B., Kınacı, A.C., Onsekiz, Üniversitesi, M., Bölümü, M.: Görüntülerdeki Araba Nesnelerinin Belirlenmesi İçin Derin Öğrenme Ile Bir Model Eğitilmesi (2018). https://developer.nvidia.com/cudnn
8. Sevimli, K.K., Deniz, M.S., Okulu, H., Bölümü, M.: Araçsal Ağlar (2010)
9. Alladi, T., Gera, B., Agrawal, A., Chamola, V., Yu, F.R.: DeepADV: a deep neural network framework for anomaly detection in VANETs. IEEE Trans. Veh. Technol. **70**, 12013–12023 (2021). https://doi.org/10.1109/TVT.2021.3113807
10. Kandali, K., Bennis, L., Bennis, H.: A new hybrid routing protocol using a modified k-means clustering algorithm and continuous hopfield network for VANET. IEEE Access **9**, 47169–47183 (2021). https://doi.org/10.1109/ACCESS.2021.3068074
11. Seo, E., Song, H.M., Kim, H.K.: GIDS: Gan based intrusion detection system for in-vehicle network. In: 16th Annual Conference on Privacy, Security and Trust (PST), pp. 1–6 (2018)
12. Naqvi, I., Chaudhary, A., Rana, A.: Intrusion detection in VANETs. In: 2021 9th International Conference on Reliability, Infocom Technologies and Optimization (Trends and Future Directions), ICRITO 2021. Institute of Electrical and Electronics Engineers Inc. (2021). https://doi.org/10.1109/ICRITO51393.2021.9596141
13. Faisal, S.M., Zaidi, T.: Timestamp based detection of Sybil attack in VANET. Int. J. Netw. Secur. **22**, 399–410 (2020). https://doi.org/10.6633/IJNS.202005
14. Kang, M.J., Kang, J.W.: Intrusion detection system using deep neural network for in-vehicle network security. PLoS ONE **11** (2016). https://doi.org/10.1371/journal.pone.0155781
15. Song, H.M., Woo, J., Kim, H.K.: In-vehicle network intrusion detection using deep convolutional neural network (2020)
16. Hussein, A., Elhajj, I.H., Chehab, A., Kayssi, A.: SDN VANETs in 5G: an architecture for resilient security services. In: 2017 Fourth International Conference on Software Defined Systems (SDS) (2017)
17. Alshammari, A., Zohdy, M.A., Debnath, D., Corser, G.: Classification approach for intrusion detection in vehicle systems. Wirel. Eng. Technol. **09**, 79–94 (2018). https://doi.org/10.4236/wet.2018.94007
18. Yang, Q., Yoo, S.J.: Hierarchical reinforcement learning-based routing algorithm with grouped RSU in urban VANETs. IEEE Trans. Intell. Transp. Syst. (2024). https://doi.org/10.1109/TITS.2024.3353258
19. Altan, G., Pasalioglu, F.: Android malware application detection using multi-layer perceptron. J. Intell. Syst. Appl. **5**(2), 95–99 (2022)
20. Muneer, S., Farooq, U., Athar, A., Ahsan Raza, M., Ghazal, T.M., Sakib, S.: A critical review of artificial intelligence based approaches in intrusion detection: a comprehensive analysis. J. Eng. **2024**(1), 3909173 (2024)
21. Altan, G.: Breast cancer diagnosis using deep belief networks on ROI images. Pamukkale University J. Eng. Sci. **28**, 286–291 (2022). https://doi.org/10.5505/pajes.2021.38668
22. Alsaade, F.W., Al-Adhaileh, M.H.: Cyber attack detection for self-driving vehicle networks using deep autoencoder algorithms. Sensors **23** (2023). https://doi.org/10.3390/s23084086
23. Martinelli, F., Mercaldo, F., Nardone, V., Santone, A.: Car hacking identification through fuzzy logic algorithms (2017)
24. Bangui, H., Ge, M., Buhnova, B.: A hybrid data-driven model for intrusion detection in VANET. In: Procedia Computer Science, vol. 184, pp. 516–523. Elsevier B.V. (2021). https://doi.org/10.1016/j.procs.2021.03.065

25. Mahmood, J., Duan, Z., Yang, Y., Wang, Q., Nebhen, J., Bhutta, M.N.M.: Security in vehicular Ad-Hoc networks: challenges and countermeasures. Secur. Commun. Netw. **2021** (2021). https://doi.org/10.1155/2021/9997771
26. Karthiga, B., Durairaj, D., Nawaz, N., Venkatasamy, T.K., Ramasamy, G., Hariharasudan, A.: Intelligent intrusion detection system for VANET using machine learning and deep learning approaches. Wirel. Commun. Mobile Comput. **2022** (2022). https://doi.org/10.1155/2022/5069104
27. Mahmood, K., Arshad, J., Chaudhry, S.A., Kumari, S.: An enhanced anonymous identity-based key agreement protocol for smart grid advanced metering infrastructure. Int. J. Commun. Syst. **32** (2019). https://doi.org/10.1002/dac.4137
28. Nayyar, A.: Flying ad hoc network (FANETs): simulation based performance comparison of routing protocols: AODV, DSDV, DSR, OLSR, AOMDV and HWMP (2018)
29. Shams, E.A., Rizaner, A., Ulusoy, A.H.: Trust aware support vector machine intrusion detection and prevention system in vehicular ad hoc networks. Comput. Secur. **78**, 245–254 (2018). https://doi.org/10.1016/j.cose.2018.06.008
30. Almi'ani, M., Ghazleh, A.A., Al-Rahayfeh, A., Razaque, A.: Intelligent intrusion detection system using clustered self organized map. In: 2018 Fifth International Conference on Software Defined Systems (SDS) (2018)
31. Nie, L., Li, Y., Kong, X.: Spatio-temporal network traffic estimation and anomaly detection based on convolutional neural network in vehicular ad-hoc networks. IEEE Access **6**, 40168–40176 (2018). https://doi.org/10.1109/ACCESS.2018.2854842
32. Kumar, R., Sharma, N.V.K., Chaurasiya, V.K.: Adaptive traffic light control using deep reinforcement learning technique. Multimed. Tools Appl. **83**, 13851–13872 (2024). https://doi.org/10.1007/s11042-023-16112-3

The manufacturer's authorised representative in the EU is Springer Nature Customer Service Centre GmbH, Europaplatz 3, 69115 Heidelberg, Germany. If you have any concerns regarding our products, please contact ProductSafety@springernature.com

Printed and bound by CPI Group (UK) Ltd, Croydon, CR0 4YY
30/09/2026
02225265-0006